REALISM FOR REALISTIC PEOPLE

In this innovative book, Hasok Chang constructs a philosophy of science for 'realistic people' interested in understanding and promoting the actual practices of inquiry in science and other knowledge-focused areas of life. Inspired by pragmatist philosophy, he reconceives the very notions of reality and truth on the basis of his concept of the 'operational coherence' of epistemic activities and offers new pragmatist conceptions of truth and reality as operational ideals achievable in actual scientific practice. Rejecting the version of scientific realism that is concerned with claiming that our theories correspond to an ultimate reality, he proposes instead an 'activist realism': a commitment to do all that we can actually do to improve our knowledge of realities. His book will appeal to scholars and students in philosophy, science and the history of science, and all who are concerned about the place of science and empirical truth in society.

HASOK CHANG is the Hans Rausing Professor of History and Philosophy of Science at the University of Cambridge. He is the author of *Inventing Temperature: Measurement and Scientific Progress* (2004), joint winner of the 2006 Lakatos Award, and of *Is Water H₂O? Evidence, Realism and Pluralism* (2012).

REALISM FOR REALISTIC PEOPLE

PEOPLE

A New Pragmatist Philosophy of Science

HASOK CHANG

University of Cambridge

CAMBRIDGE
UNIVERSITY PRESS

CAMBRIDGE
UNIVERSITY PRESS

Shaftesbury Road, Cambridge CB2 8EA, United Kingdom

One Liberty Plaza, 20th Floor, New York, NY 10006, USA

477 Williamstown Road, Port Melbourne, VIC 3207, Australia

314–321, 3rd Floor, Plot 3, Splendor Forum, Jasola District Centre, New Delhi – 110025, India

103 Penang Road, #05–06/07, Visioncrest Commercial, Singapore 238467

Cambridge University Press is part of Cambridge University Press & Assessment, a department of the University of Cambridge.

We share the University's mission to contribute to society through the pursuit of education, learning and research at the highest international levels of excellence.

www.cambridge.org
Information on this title: www.cambridge.org/9781108455930

DOI: 10.1017/9781108635738

First published 2022
First paperback edition 2023

A catalogue record for this publication is available from the British Library

ISBN   978-1-108-47038-4   Hardback
ISBN   978-1-108-45593-0   Paperback

To Gretchen Siglar,
for her curiosity, her insight and her love

Contents

List of Illustrations   *page* ix
Acknowledgements   x

Introduction   1
    What Is the Problem?   1
    Pragmatist Notions of Knowledge, Truth and Reality   3
    Scientific Realism as Realistic Activism   6
    What Kind of Book Is This?   8
    A Note on Structure   11

1 Active Knowledge   12
    1.1 Overview   12
    1.2 Epistemic Agents   27
    1.3 Epistemic Activities and Systems of Practice   33
    1.4 Operational Coherence   40
    1.5 Inquiry as Aim-Oriented Adjustment   48
    1.6 Pragmatism and Active Knowledge   57

2 Correspondence   68
    2.1 Overview   68
    2.2 Correspondence Realism: Between Metaphysics and Semantics   83
    2.3 Can Reference Save Correspondence Realism?   92
    2.4 Faith in Science   100
    2.5 Real Representations   111

3 Reality   119
    3.1 Overview   119
    3.2 How Mind-Framing Works   131
    3.3 The Achievement of Reality   142
    3.4 Ontological Pluralism   148
    3.5 Putting Things Together   155

4 Truth   163
   4.1 Overview   163
   4.2 Different Kinds of Truth   173
   4.3 Empirical Truth and Operational Coherence   179
   4.4 Truth as a Quality   186
   4.5 Plurality and Incommensurability   192
   4.6 Rehabilitating the Pragmatists   197

5 Realism   204
   5.1 Overview   204
   5.2 Pragmatism and Realism   216
   5.3 Internal and Perspectival Realism   223
   5.4 Pluralism and Realism   231
   5.5 Epistemic Iteration Revisited   239
   5.6 Progress and the Scientific Realism Debate   247

Closing Remarks   252
   A Humanist Vision of Knowledge   252
   The Road Ahead   253
   Bringing Philosophy Back to Life   254

Bibliography   256
Index   275

Illustrations

1.1 An experimental arrangement demonstrating the
electromagnetic effect first discovered by
Hans Christian Ørsted   *page* 52
1.2 Faraday's compact rotation device   55
2.1 Ball-and-stick molecular models by August Hofmann   112

ix

Acknowledgements

So many people have helped in the making of this book that I will not be able to list anywhere near all of them here. I will do my best to note my most salient debts, with apologies and thanks to all those who will have to go unnamed.

First of all I want to thank all of my family members near and far for being there for me, especially my parents back in Seoul, whose steadfast love and support never will wane.

I could never have completed this project without the privilege of a three-year research leave that I enjoyed as a British Academy Wolfson Research Professor. Being selected for this honour also gave me the assurance that this project was worth doing. I thank the British Academy and the Wolfson Foundation most sincerely.

I would like to thank all the people at Cambridge University Press who helped bring this book out into this world. Most of all I thank Hilary Gaskin for her long-term encouragement and support, and patience with my changing plans and ever-slipping schedule. I would also like to thank Neena Maheen and colleagues at Straive for the production work. And Damian Love's expert and meticulous attention in copyediting has been a true blessing.

This book could not at all have become what it is without the generous help of Philip Kitcher, who encouraged and instructed me in this direction of work for many years. Crucially, he helped me gain the British Academy award, reviewed the proposal for this book and then the full manuscript for the press, and gave me what I will always think of as the shining exemplar of constructive criticism. And he did it all without ever having any obligation to mentor me. In a very similar vein I would also like to record my deep thanks to Martin Carrier.

Equally indispensable has been the help from my 'Angels' (PhD students and others), who worked as my 'pacemakers' as I raced to make the first full draft of this book, and gave me endless advice and encouragement.

Those weekly Zoom meetings through the coronavirus pandemic will live forever in my fond memory. I thank especially Céline Henne, Oscar Westerblad, Helene Scott-Fordsmand, Emilie Skulberg and Bob Vos; equally insightful feedback, though less extensive, came from Sarah Hijmans, Milena Ivanova, Hannah Tomczyk, Agnes Bolinska, and Grace Field. I also thank Daniel Ott, Rory Kent and Henrique Gomes for their contributions. The Angels also continue to be my own European Union through the dark days of Brexit.

Similarly, at earlier stages of work, groups at some leading centres of philosophy of science have provided memorable formative occasions for my work on this project. I will always remember with great appreciation the seminars and reading groups at Bielefeld (hosted by Martin Carrier), Tartu (Ave Mets and Endla Lõhkivi), Edinburgh (Michela Massimi) and UCL/LSE (Chiara Ambrosio and Roman Frigg). Also important have been occasions to present my work at the Aristotelian Society (Hannah Carnegy, Guy Longworth), Leeds (Greg Radick), Hanyang University (Sang Wook Yi), Vienna (Martin Kusch, Fritz Stadler, Elisabeth Nemeth), Ghent (Erik Weber), Oxford (Simon Saunders, Harvey Brown), TU-Berlin (Friedrich Steinle), Pardubice (Filip Grygar), Pittsburgh (Sandy Mitchell), LOGOS and UAB in Barcelona (Carl Hoefer, Genoveva Martí, Thomas Stürm and Silvia de Bianchi), Bern (Andrea Loettgers), Simon Fraser (Holly Andersen), Toronto (Hakob Barseghyan), Montevideo (Lucía Lewowicz) and many other places, as well as various conference series including SPSP, &HPS, PSA, CLMPST and the European Pragmatism Conference. I thank not only the people named above as my main hosts, but many others who lent me their thoughts and ears on all those occasions.

There are a number of other people that I would like to thank for their helpful critical and constructive feedback on earlier versions of various parts of this book over many years. It is impossible to mention all of them, but I especially want to mention Nancy Cartwright, Roberto Torretti, Paul Teller, Miriam Solomon, Léna Soler, Jamie Shaw, Mike Stuart, Alison Wylie, Sabina Leonelli, Jeremy Butterfield; and also Anna Alexandrova, Rachel Ankeny, Theodore Arabatzis, Marina Banchetti, Ann-Sophie Barwich, Mieke Boon, Karim Bschir, Julia Bursten, Anjan Chakravartty, Alan Chalmers, Mazviita Chirimuuta, Stijn Conix, Claudia Cristalli, Adrian Currie, Henk de Regt, Stuart Firestein, Dagfinn Føllesdal, Peter Galison, Marta Halina, Clevis Headley, Robin Hendry, Gerald Holton, Phyllis Illari, Alistair Isaac, Katie Kendig, Ken Kendler, Helen Lauer, Tim Lewens, Geoffrey Lloyd, Cheryl Misak, Miguel Ohnesorge, Themis

Pantazakos, Lydia Patton, Huw Price, Karoliina Pulkkinen, Ken Schaffner, Dunja Šešelja, William Simpson, George Smith, Christian Straßer, Mauricio Suárez, Katie Tabb, Dave Teplow, Nick Teh, Peter Vickers, Rick Welch, Bill Wimsatt and Jim Woodward.

Nearer home, I thank my colleagues, students and other members and associates of the Department of History and Philosophy of Science at the University of Cambridge for providing a most congenial and stimulating intellectual environment and first-rate library resources and administrative support.

Through all these interactions and many others, I have felt that a certain global scholarly village felt the need for a book like this, and somehow nominated me to write it. In actual practice this meant that when I presented earlier versions of the material contained in this book, so many of you encouraged me so enthusiastically but also told me that I needed to do more, and better. I especially have in mind the community that formed through the Society for Philosophy of Science in Practice (SPSP). This book is for you, and I hope it is worth the wait.

Introduction

What Is the Problem?

The aim of this book is to develop a realistic philosophical vision of scientific knowledge, which can be truly relevant to scientific work and useful in social and political life. In a world filled with so much misinformation, ignorance, prejudice, deception and mistrust, where can we turn for reliable facts, insightful theories, and guidelines for action? I am old-fashioned enough to believe that science and the scientific attitude constitute our best hope. But we are often distracted by an impossible ideal of scientific knowledge as proven universal truth about some ultimate reality. Unrealistic ideals can have harmful consequences. This is quite evident in the current crisis in the public's trust in science. When the advocates of science imply that science is the giver of unquestionable truth, a false equivalence opens up when actual science fails to live up to its overblown image. People will think: if evolution is not completely proven, it is 'just a theory' to be treated equally with creationism. They will think: if it is not completely proven that human activity is causing climate change, and moreover some scientists disagree with the idea, then it is just a majority *opinion* among scientists, not inherently superior to *their* own opinion.

Faced with challenges to established scientific knowledge, many scientists, philosophers and concerned citizens succumb to an authoritarian or even autocratic impulse, which they often honour with the label of 'scientific realism': science gives us the Truth about Reality (or at least an approximation to it), and anyone going against the verdict of science is simply misguided (if not malicious). There is a widespread conceit that modern science has *basically* the right answers, or at least the right *methods* for getting the right answers. Many 'realists' maintain this idea like an article of faith, even though we have no direct access to the unobservable aspects of Reality and the historical track-record of science shows serious fluctuations in scientists' views about the most basic aspects of the universe. In the philosophy of science, what usually goes by the name of (scientific) realism comes down to the notion that our best scientific theories must express some version of the final answers to the most fundamental questions about nature. I cannot help expressing the feeling that this widespread philosophy of scientific realism amounts to an appropriation of the term 'realism' to describe a most *unrealistic* doctrine. I want to propose a realistic kind of realism, close to what William Wimsatt has called 'realism for limited beings in a rich, messy world', which is 'a philosophy of science that can be pursued by *real* people in *real* situations in *real* time with the kinds of tools we *actually* have' (Wimsatt 2007, p. 5; emphases original).

One problem with the standard doctrine of 'scientific realism' is that it tells us very little about what we can *do* to improve scientific knowledge. We need an *operational* ideal of knowledge, one that we can actually work with, which reflects our best actual practices while also indicating how they can be improved further. Ideals are of course hardly ever achievable, but in order to serve a useful function an ideal needs to be something that has an effect of making us think and behave differently. Inventing a test for a virus with zero per cent rate of false negatives, or having no murders in a large city in a whole year, is an operational ideal, even if we never manage to meet it. Seeking absolute truth is not an operational ideal – there is nothing we can actually do in order to approach that ideal. According to the common picture of scientific knowledge, science should give us the true picture of the reality that exists well-formed 'out there' completely independently of our conceptions and our experiences. But such 'reality' is not accessible to us and there are no actual methods by which we could attain assured knowledge about it.

A realistic ideal of knowledge also recognizes that a mature science in a democratic society needs to allow debate, treating dissenters with respect in the spirit of what Philip Kitcher has called the 'ideal conversation' (2011a, ch. 2).[1] Even if effective practical action necessitates the overriding of certain views and the sidelining of certain voices, this should still be done with intellectual sympathy and compassion. Behaving like religious fundamentalists is not the way to beat religious fundamentalism. We must overcome the urge simply to shut down the opposition in science, just as the most mature societies have learned to do in politics. We should not look to science for the authority to justify repression. Science consists in

---

[1] Philip Kitcher and Evelyn Fox Keller (2017) have concretely imagined such ideal conversations concerning climate change.

continual inquiry and debate, not a collection of dogmas that are protected from criticism. As my childhood hero Carl Sagan said about science: 'Its only sacred truth is that there are no sacred truths.'[2] Even though some dogmas are sometimes practically necessary, dogmatism should not be our way of life.

You might wish that scientists themselves would set up and maintain a framework for democratic debate about scientific knowledge. I believe that some scientists did make such efforts back in the Enlightenment era, after the initial excitement of rebelling against traditional doctrines passed and science needed to look more openly and humbly at its own growing epistemic authority (see Chang 2012b). But a relentless process of professionalization has left many modern scientists oblivious to such needs. In that situation, it is appropriate that philosophers of science should take up the challenge. But the sad reality is that much traditional philosophical thinking has been caught up in the contemplation of the impossible ideal of knowledge mentioned above, and failed to provide insights that are relevant to actual practices of science. Even though much good work has been done in various corners of the 'philosophy of the special sciences' in communication with scientific practitioners, the spirit of such work has failed to permeate the discourse in general philosophy of science, not to mention epistemology and metaphysics done without particular attention to science. Among those who care about the place of science in society and culture, there is an urgently felt need for a general philosophy of science fit for engaging with real scientific practices. That is the need that this book seeks to meet.

Pragmatist Notions of Knowledge, Truth and Reality

Here is the most fundamental step to take, in order to bring our philosophical views of scientific knowledge more in line with real practices: we should be thinking about what it is that we *do* in science. I find inspiration in the operationalist philosophy of Percy Bridgman (1956, p. 76): 'it is better, because it takes us further, to analyze into doings or happenings rather than into objects or entities.'[3] This proposal will take us clearly beyond the default instinct of analytic philosophers to regard knowledge as

---

[2] This statement occurs in the PBS Television series *Cosmos*, episode 13. The equivalent passage in the book can be found in Sagan (1980, p. 333): 'It [science] has two rules. First: there are no sacred truths.'

[3] Bridgman's thoughts were often misunderstood, and he did not like the label of 'operationalism'. See Chang (2009b; 2017c) for my interpretation.

consisting of justified belief in a set of true statements or propositions (or something of that sort). As I will explain in Chapter 1, the alternative perspective I promote takes science (and inquiry in general) as something that people do, consisting of epistemic activities with various aims whose achievement we can actually assess (unlike absolute truth). From this perspective, knowledge will be seen primarily as an ability, not confined to the possession of information. Having reliable beliefs is one aspect of the ability to conduct successful activities, but it is by no means the whole picture.

In this broadly operationalist spirit I propose that we rethink the philosophical notions of 'truth' and 'reality', so that they can be used without encouraging unrealistic and indefensible dogmatism. Truth and reality are perfectly meaningful concepts in the actual business of 'representing and intervening' (Hacking 1983), and they should stay in that realm of practice. We need to be able to debate the truth and falsity of statements, without claiming our truths to be absolute. We need to be able to say things like 'The President is out of touch with reality on this point', without in the same breath invoking some realm of metaphysical reality that we cannot ourselves claim to be in touch with. Articulating such useful notions of truth and reality, and a practical ideal of knowledge based on them, is something that I regard as one of the most important tasks of philosophy.

My proposed reframing of knowledge, truth and reality is built on the notion of 'operational coherence', which will be spelled out in Chapter 1. Very roughly, operational coherence is a matter of making elements of our activities fit together harmoniously so that our aims may be achieved. It may consist in something as mundane as the skilful coordination of bodily movements and material conditions in riding a bicycle, or it may be something as esoteric as the successful integration of material technologies and abstract theories in the operation of the global positioning system (GPS). Building on that notion of coherence, I advance a notion of truth in Chapter 4 according to which the empirical truth of a statement consists in the positive role it can play in facilitating operationally coherent activities. Similarly, in Chapter 3 I propose that what we should mean by something being 'real' is that it can be employed in coherent activities that rely on its existence and its basic properties. Truth and reality conceived in such a way are attributes grounded in our activities.

In terms of philosophical traditions, what I want to do is to reframe the basic discourse in the philosophy of science through a revitalization of pragmatism.[4] Among the established philosophical frameworks, pragmatism offers us the best hope for facilitating proper attention to what it is that we really do when we gain and use knowledge. Pragmatism regards knowledge as an outcome of humble on-the-ground inquiry, and locates it in actual intelligent activities we carry out in life. I hasten to add that pragmatism has often been misrepresented and vilified within philosophy, so much so that many people who hold what I would consider pragmatist views have avoided the label of pragmatism. Even Marjorie Grene, who made a pioneering emphasis on how knowledge exists in the context of action, shrank sharply from the label of pragmatism, which she took as an insult even worse than relativism: 'But pragmatism! If that is the kind of abandonment of any interest in cognitive claims we are supposed to be advocating, we are presenting our case very badly' (Grene 1987, p. 69). There may be times when a word is so soiled by negative connotations that we should just give it up instead of engaging in futile attempts to correct the misuses and abuses of it. But I think the term 'pragmatism' can and should be defended and improved, not given up – as with 'truth' and 'reality'.

In later parts of the book I will give a more considered and systematic view of what I take pragmatism to mean, but it would be helpful to start with a few brief notes of clarification. The common misrepresentation of pragmatism comes from two different directions. First, there is the charge that pragmatists disregard truth in any meaningful sense of the word, because pragmatism allows one to regard as truth whatever is *convenient* to oneself. Nothing could be a worse distortion of pragmatism. Whether our conceptions work out in practice is most definitely not up to our whims and wishes, nor is it a matter of what is 'convenient' to us in a shallow and limited sense. In order to do things coherently, we need to have a mastery of our surroundings. It is actually through operational coherence, not by the mirage of correspondence to the inaccessible 'Real World', that something objective is brought to bear on our knowledge.

In the second common line of misrepresentation, it is alleged that pragmatism concerns itself only with 'practical' things, and people have even blamed it for the currently prevalent devaluing of the humanities and other 'impractical' fields of study. True pragmatism is about how our conceptions work out in all practices, not just in the money-making or enemy-killing sorts of 'practical' practices. There is huge variety within human practices, in pursuit of diverse types of aims, and pragmatism is

---

[4] In many ways I follow the directions indicated by James Woodward (forthcoming).

concerned with all of them. For example, most of the experiments going on at CERN and other places of research in pure science are decidedly not 'practical', at least not immediately. Yet what goes on at such installations is the epitome of the empirical testing of theories, which pragmatists admire and cherish. The pragmatist vindication of the concept of imaginary number is in all the fruitful mathematical practices it enables, including practically useless ones as well as engineering applications. The pragmatist vindication of the fugue is in the experience of elation that countless listeners have felt in hearing Bach's masterpieces; whether such musical pieces can be put to 'practical' uses such as helping soldiers march in step is irrelevant to their pragmatist appreciation.

What the spirit of pragmatism recommends is that philosophy should not be removed from the various practices in life, including science. What philosophers need to do is think different and unusual thoughts, but also about things that matter to people here and now. We do need the proverbial armchair in which to do philosophy, but only as a place of necessary and occasional retreat to allow us to think quietly and carefully about issues of actual life. It may be surprising to some to learn that such practical engagement was an inclination strongly expressed in the mani-festo of logical positivism by the Vienna Circle (which I quote with apologies for the gendered language of the time):

> Neatness and clarity are striven for, and dark distances and unfathomable depths rejected. In science there are no 'depths'; there is surface everywhere ... Everything is accessible to man; and man is the measure of all things. Here is an affinity with the Sophists, not with the Platonists; with the Epicureans, not with the Pythagoreans; with all those who stand for earthly being and the here and now. (Neurath et al. [1929] 1973, p. 306)

Scientific Realism as Realistic Activism

The philosophical battleground most relevant to my concerns expressed above is the fierce and long-standing debate on scientific realism. Ilkka Niiniluoto opens his erudite and insightful book on scientific realism thus: 'The philosophy of science in the twentieth century has been a battlefield between "realist" and "anti-realist" approaches' (Niiniluoto 1999, p. v). I wish to take philosophers' attention away from standard scientific realist attempts to show that the impossible is *somehow* possible, namely that empirical science can attain assured truths about what truly goes beyond experience. As I will argue in Chapter 2, it is time to accept the fact that we cannot know whether we have got the Truth about the World (and that such thoughts are perhaps not even meaningful). Scientific realists go astray by persisting in trying to find a way around this fact, while anti-realists make the mistake of engaging unproductively with that realist persistence.

I would like to reshape realism into a stance[5] that is useful for scientists and others who are actually engaged in the production, assessment and improvement of empirical knowledge: a *realism for realistic people*, which will be fully articulated in Chapter 5. I am trying to carry out a task that I already advocated a decade ago: 'I think *realistic* people (including most empiricists and pragmatists) should re-claim the label of "realism"!' (Chang 2012a, p. 217; emphasis original). This is a call that had already been made by many others. For example, there is some affinity between my line of thought and what Peter Kosso (1998, ch. 8, pp. 177–8) called 'realistic realism'. I cannot remember whether I absorbed the 'realistic' trope from Kosso, but I would be happy if that were the case; in Section 5.1 I will also mention various other authors who have voiced a realistic spirit.

I hope that my articulation of an operational ideal of knowledge in this book will persuade both traditional 'realists' and 'anti-realists' that there is a more realistic version of realism that they can all subscribe to. The realistic realism I offer here is focused on genuine empirical learning, and provides a notion of knowledge suitable for conceptualizing how rational decision-making works in real life. Knowledge is only meaningful within the world in which we live. It is a futile and pernicious philosophical dream to seek what Hilary Putnam (1980a, p. 100; 1981, p. 49) called the 'God's Eye point of view' on nature, an 'externalist' perspective from which we can tell the 'real' shape of the world. Roberto Torretti (2000, p. 114) blasts the 'scientific realists' who believe 'that reality is well-defined, once and for all, independently of human action and human thought' – yet 'in a way that can be adequately articulated in human discourse'. The self-designated 'realists' hold that science aims to develop 'just the sort of discourse which adequately articulates reality – which, as Plato said, "cuts it at its joints" – and that modern science is visibly approaching the fulfilment of this aim'. Torretti confesses that he finds it difficult 'to accept any of these statements or even to make sense of them'. Epistemology should not be focused on the non-existent 'final' state of knowledge. And it is futile to insist that

---

[5] I use the term 'stance' in a similar spirit as in van Fraassen (2002) on empiricism and Kellert, Longino and Waters (2006) on pluralism. Here is a slightly later formulation from van Fraassen (2004, p. 128): 'A stance consists of a cluster of attitudes, including propositional attitudes (which may include some factual beliefs) as well as others, and especially certain intentions, commitments, and values.'

a vision of scientific knowledge expressed from the God's Eye view is useful as a regulative ideal; nothing can be considered 'regulative' if it has no actual bearing on how we conduct ourselves, if it doesn't in fact help *regulate* our practices.

The realism I advocate in this book is an *activist* stance, as I will explain further in Chapter 5. When it is applied to science, it is the philosophical position that says science should do everything it can in order to gain more and better knowledge, as opposed to the position of a spectator who observes with satisfaction that science seems very good at finding knowledge, and that this knowledge must be a fair reflection of the state of the real world. Most scientists would clearly endorse the activist stance if questioned about it, though their actual conduct sometimes goes in a different direction. It is worth noting that the activist nature of the realism I advocate makes it a very different kind of doctrine from metaphysical realism (with which 'scientific realism' usually aligns itself, as discussed in Chapter 2). It makes sense that the pronouncements of metaphysical realism ('The external world is real!') leaves most practising scientists cold ('Go away, I'm busy learning something about nature!'). Realism as I see it is a stance in active service of scientific progress. It applies in a similar spirit to other areas of life, too, wherever we regard knowledge as a good thing. It is an injunction against resting content with what we believe, or what we think we know. Such a realist spirit is intimately connected to social and political progress as well.

What Kind of Book Is This?

Many philosophers have an ambition to write a book that becomes the focus of philosophical debates everywhere. My ambition is focused more on writing a useful book. What I am trying to offer is a set of ideas that people in various walks of life can pick up as tools for their own work, ideas whose worth will hopefully be proven through productive use by those who want to understand and promote good practices relating to knowledge. In my own research, the immediate use of ideas in the philosophy of science has always been in the framing of the history of science. The ideas in this book are being applied to (and shaped by) some new historical research I am conducting at the same time, whose results will be published in a separate book (*How Does a Battery Work?*, forthcoming from the University of Chicago Press). I hope that the philosophical ideas developed in the present book will provide useful framing devices for other historians, too, as well as sociologists, anthropologists and others who make empirical studies of science. I also hope that my ideas can help practitioners of science who want to make more explicit sense of the aims and methods of the work that they do so well. When I say 'practitioners of science' I have a broad range of people in mind: not only research scientists, but engineers, doctors, mathematicians, and teachers and students of all those fields, too. Even more broadly, I hope that this book will be useful to all those who want to think carefully about the place of knowledge in our individual and social lives, including policy-makers and the general public. The operational ideal of good knowledge that I am proposing here will hopefully aid clear thinking in all the areas where science meets politics and ethics.

I do not focus on the adversarial type of argumentation, except where I feel that it is really necessary for creating the space in which my own views can grow, as in Chapter 2. I will tend to bypass many of the ongoing cutting-edge debates in relevant areas of professional philosophy. This is not from a lack of respect for that work, but simply a matter of my own priorities. It is not my main concern to argue that other people have been wrong, especially when I know that many of them are superior intellects to myself who have put a great deal of effort into developing and defending their views. In fact, avoiding excessive and minute disputation is part of my aim. I do not wish to write the kind of specialized academic philosophy that is practically unusable even for most professional philosophers not working in the immediate narrow sub-fields within which the debates take place.

I am most interested in calling for a change of perspective or stance, rather than proving specific points. Sometimes I will merely be introducing useful tautologies, for example when I say that 'all we can ever think or speak about are conceptualized entities' (see Chapter 3).[6] I aspire to present an overall vision that you, Dear Reader, can subscribe to, even if you disagree with some particular points I make. And, in any case, I do not think that philosophy is the kind of enterprise in which one can prove points absolutely or win arguments decisively. I am passionate about the ideas I am presenting, but I seek to master the art of *respectful* denunciation, *peaceful* incitement, and *productive* frustration. I would like to be reasonable and rational without going on about Reason and Rationality, to be realistic without grand claims about Reality, and to be honest and true without banging on the table about Truth.

---

[6] Everyday examples of useful tautologies include 'A man's gotta do what he's gotta do' and the recently infamous 'Brexit means Brexit' (Theresa May, UK prime minister).

This is the first time I have ever attempted to write an entire book of abstract philosophy. In departing from my normal mode of work I felt that I was answering a call of duty, though the work has certainly been pleasurable, too. The ideas I am presenting here needed to be articulated, and I could not find them already put together in a clear, systematic and accessible way anywhere. So I was compelled to try to articulate them in my own way, for myself and for others. I am not even good at sustained abstract thinking, especially compared to the formidable thinkers who are the leading lights of academic philosophy, whose works justly inspire awe and admiration. But I take comfort in the thought that geniuses are rarely good teachers or explainers. And I did not in fact succeed in writing an entirely abstract book. On the contrary, the discussions to follow will be peppered with many concrete examples from the history of science, as well as everyday life, though there are no sustained historical studies. One caveat: you will find that most of my scientific examples are from physics and chemistry, not drawn in a balanced way from across the natural, social and human sciences. To be frank, this is because physics and chemistry are the sciences I know best. But I think my limitation can also be an advantage, because the kind of traditional philosophical views about science that I am trying to move beyond have largely been inspired by the physical sciences, especially physics. By including many examples from physics as illustrations of my points, I am making the point that even physics isn't like 'physics' (i.e., the common image of physics).

It is my own honest view that there are very few truly original ideas contained in this book. While I don't agree with the common adage that all philosophy is footnotes to Plato and Aristotle, I do think that it is difficult to have a completely original thought on any important issue that people have worried about for centuries. My thoughts have been inspired by a number of great thinkers past and present, ranging from Immanuel Kant to Nancy Cartwright. What I do think is original in my book is the particular way in which I synthesize the old familiar thoughts, which I hope will also enhance the value and meaningfulness of each component going into the synthesis. Again, my main aim is to present a positive view that you can use for your own purposes. I will make my best attempt to connect and engage with other authors' works as needed, but in general my focus will not be on fine-grained disputation or exegesis. Rather, I honour other thinkers by using their ideas as inspirations in thinking for myself, and by building directly on their work whenever I can. And I hope that you will do the same with my work.

A Note on Structure

This book has a multi-layered branching structure, which is meant to make it as accessible as possible. The structure is also intended to assist effective and coherent selective reading, accepting the reality that most readers will not read every word of it cover-to-cover. In this Introduction I have begun with a very brief and intuitive overview of the whole. This will now be followed by a set of chapters that enter into details on particular aspects of the story. That is the overall trunk-and-branch structure of the book.

At the next level, each chapter will also have a branching structure: an overview section followed by a set of more specialized sections, which enter into fuller technical details on particular issues, often anticipating specific objections or demands for elaboration. Each of these specialized sections also begins with a brief summary, followed by the details. In principle, the discourse could continue indefinitely into more specificity in a fractal-like fashion. In order to indicate the levels visibly, this Introduction, the opening overview section of each chapter, and the beginning summaries of the other sections are printed in a different typeface from the rest of the text.

This branching structure should work more like a well-constructed newspaper article or website than a mathematical proof, though there will be some deductive arguments contained in specific parts of the book. You can take this book as you wish – short or long, succinct or elaborate. Just read the top-level text (in this typeface) if you are busy, or not *that* interested. If you intuitively get and agree with the overview of a given chapter or section, you may not want to read any further in it. But I hope you will want to read on, at least down some particular branches. The branching structure expresses my sense that inquiry can and should multiply itself continually; whatever conclusion we reach should call forth new questions and issues for further investigation.

CHAPTER I

Active Knowledge

1.1 Overview

How Science Goes beyond Propositional Knowledge

The most fundamental step in making an operational ideal of knowledge is to understand knowledge in terms of what people *do*. I want to focus on the sense of knowing how to do something, a sense of knowledge as ability. Philosophers normally take knowledge as the mental possession of true *propositions*, or *theories* (understood as organized sets of propositions). This conception is at the core of the dominant textbook treatment of epistemology with which mainstream analytic philosophers still start their training. Writing for the authoritative *Stanford Encyclopedia of Philosophy*, Jonathan Jenkins Ichikawa and Matthias Steup (2018) declare: 'the project of analysing knowledge is to state conditions that are individually necessary and jointly sufficient for propositional knowledge'. And then they remind us of the traditional ('tripartite') analysis of knowledge: 'justified, true belief is necessary and sufficient for knowledge.'[1] Of course, epistemologists at the cutting edge of research have moved far beyond this 'Epistemology 101' conception. But my own feeling is as with the old joke: 'How do I get to Letterfrack?' – 'I wouldn't start from here.'

My motivations for looking for a different starting-point come chiefly from the needs of the philosophy of science. The propositional view is simply not commodious enough for understanding scientific *practices*. And even for those who are not philosophers of science, paying attention to science in making a theory of knowledge makes good sense. While most aspects of human life involve knowledge in some way, in science the acquisition of knowledge is our main aim. Therefore, by observing what

---

[1] Similar statements are too numerous to cite in any comprehensive way. For typical examples see Dancy (1985, p. 23), and Audi (2014, p. 221).

people do in science we can learn something about what is involved in knowledge acquisition with relatively little background noise, and it is likely that we can learn some lessons applicable to knowledge in non-scientific situations as well.

Let us start by observing, without too much initial prejudice, what it is that scientists do. What kinds of things do we want to know in science? We do, of course, want to know some cut-and-dried facts. But we also want explanation and understanding, which many lovers of science prize above all. If philosophers don't want to deal with anything fuzzy like understanding, how about the knowledge of causes and mechanisms, at least? And what about having a sense of what it's like to experience nature, whether it be to smell chlorine gas or to communicate with a chimpanzee? What about knowing how to create new entities and phenomena, such as nuclear fission chain-reactions, or new 'super-heavy' chemical elements? Aren't all these things in the realm of scientific knowledge? It is not obvious at all that the main focus of scientific knowledge-making is just on propositions or theories, but we don't seem to have a philosophically rigorous way of thinking about this diverse array of types of scientific knowledge. Here we need to improve philosophy, instead of trying to dismiss or explain away the bits of science that do not fit our standard philosophical way of thinking. Before going on to the positive task, let me outline more carefully three different ways in which the philosophy of science is hampered by an exclusive commitment to the propositional view of knowledge.

(1)   Much of what is prized as knowledge in science is in the realm of knowing how to *do something*, which is not easy to accommodate within the propositional view. As Ian Hacking (1983, ch. 13) has emphasized, *making* is an important mode of scientific work, regardless of any practical uses that our creations might have. More broadly, here is an important insight from Gilbert Ryle (1946, p. 15): 'The advance of knowledge does not consist only in the accumulation of discovered truths, but also and chiefly in the cumulative mastery of methods.' A handful of concrete examples from the rich tapestry of scientific progress in the last few centuries will give us a sense of the range of things that an epistemology of science ought to be able to handle. (I would ask you to spare one minute to take these examples in, rather than skipping through them as details unworthy of philosophical attention.) Since the eighteenth century we have known how to compute the precise trajectory of a planet, which involves

knowing how to solve the relevant basic equations of physics. It is also important to know how to come up with such equations in the first place (and invent an appropriate kind of mathematics in which they can be framed). Nowadays we also know how to run simulations of experiments that we can't carry out physically, and how to make formal models of complex situations. We know how to predict the weather in the short term, and make good prognoses for many diseases. We know how to measure all sorts of things ranging from temperature and humidity to the rate of inflation. Feats of observation are not limited to quantitative measurement, either. We have learned how to ascertain molecular compositions and structures, and complex tasks like the sequencing of DNA molecules are now routine. We know how to image extremely distant objects with devices like the Hubble space telescope. We also know how to classify things in useful and effective ways, as with biological taxonomy or the periodic table of elements. We also have a great variety of abilities in the realm of making: synthesizing pharmaceutical agents, making high-temperature superconductors, creating and operating complex technological systems such as radio, television and the internet, and making artificial intelligence that can outperform humans in many complex mental tasks.

(2) Many thinkers who have taken a serious historical look at science have come to doubt that theories were the main units of scientific knowledge. Philosophers of science had traditionally worried about the problem of *theory*-choice, but Thomas Kuhn ([1962] 1970) showed quite convincingly that scientists' choice at the most crucial moments in history tends to be a choice concerning larger and more complex wholes, which he called *paradigms*. What exactly Kuhn meant by 'paradigm' is famously debatable, but it can be readily agreed that a paradigm contains much more than descriptive statements. That is especially evident if we take Kuhn's broader sense of paradigm as 'disciplinary matrix' (rather than as 'exemplar') (Kuhn [1962] 1970, pp. 181–7). A paradigm in this sense specifies some particular *problems* deemed worth addressing, the right *methods* of tackling those problems, and the *criteria* by which the solutions to the problems are assessed. *None of these elements consists of descriptive propositions.* Larry Laudan (1977; 1984) critiqued Kuhn on the 'holist' assumption that all elements of a paradigm must change together, but he agreed that a 'research tradition' contains diverse types of elements within it: theory, methodology and axiology.

A more recent comparison would be Paul Teller's (2021, p. S5016) rendition of the concept of 'framework', or Rachel Ankeny and Sabina Leonelli's (2016) notion of 'repertoires' in scientific practice, which pays due attention to the social organization of research. All of these thinkers have stressed that theory-choice is inextricably linked to aspects of scientific practice that are not reducible to descriptive statements.

(3) The propositional view of knowledge highlights the *products* of inquiry, mostly neglecting the *processes* of inquiry. One enduring shortcoming of mainstream epistemology and philosophy of science is traceable to this issue: we distinguish the context of discovery and the context of justification, and then proceed to ignore the former. Justification is a process that can be conceived reasonably well in propositional terms, so it receives far more attention in analytic philosophy than other processes involved in inquiry, such as hypothesis-generation and concept-formation. But even if we are just dealing with justification, serious attention should be paid to the ins and outs of the cognitive processes involved in it. Most epistemologists continue to conceive justification in propositional terms, ultimately based on how well a proposition agrees with other, better-established propositions. There are mainstream epistemologists who do emphasize processes, and chief among them are the reliabilists, in a tradition reaching back to Frank P. Ramsey. However, the 'reliability' of a process, rule or agent of inquiry seems to get defined inevitably in terms of the truth of the propositions that it produces or sanctions, so we return again to the focus on products (see Goldman and Beddor 2016 and references therein). The neglect of processes creates the misleading impression that knowledge is something we passively accept or reject, rather than actively seek, create and evaluate.

We need a fresh framework for thinking about what makes a good scientific practice. In previous publications I have proposed analysing scientific work in terms of **epistemic activities**[2] (and **systems of practice**, in which various epistemic activities function jointly), in conscious opposition to the more customary analysis framed in terms of propositions (Chang 2011a; 2011b; 2012a; 2014). These concepts will be elaborated further in Section 1.3. Very briefly and intuitively for now: an activity is a programme of action designed for the achievement of a recognizable aim

---

[2] I will put in boldface key terms of my own devising on their first occurrence.

(or, the execution of such a programme); an *epistemic* activity is a knowledge-related activity, aimed at acquiring, assessing or using knowledge. A system of practice is a network of activities that function coherently together. Activities are not reducible to propositions in any straightforward sense; instead, I will be thinking about how propositions fit into activities.

As an initial illustration of an epistemic activity, let's actually take something that might seem very far removed from actions: the definition of a concept. Consider what one has to *do* in order to define a scientific term: formulate formal conditions for its correct uses; construct physical instruments and procedures for measurement, standard tests and other manipulations; round people up on a committee to monitor the agreed uses of the concept, and devise methods to punish people who do not adhere to the agreed uses. 'One meter' or 'one kilogram' would not and could not mean what it means without a whole variety of epistemic actions coordinated by the International Bureau of Weights and Measures in Paris. Even semantics is a matter of doing, as Percy Bridgman (1927) and Ludwig Wittgenstein (1953) taught long ago.

For an illustration of a system of practice, consider the new system of chemistry that Antoine Lavoisier and his colleagues created in the late eighteenth century, bringing about the so-called Chemical Revolution (Chang 2012a, ch. 1). The main epistemic activities in Lavoisier's 'oxygenist' system of practice included: making various chemical reactions; collecting the products of reactions, especially gases; identifying various substances through standard chemical tests; analysing organic substances through combustion; measuring the weights of the ingredients and products of reactions; tracking chemical substances through those weight-measurements; and so on. Some of these were already well-established activities in chemistry, and others were more novel. These activities were coordinated together for the purpose of achieving the overall aims of the system, such as the knowledge of the chemical composition of various substances, a good classification of all chemical substances, and the explanation of chemical reactions. Systems of practice form coherent bodies of scientific work, and they also make useful units of analysis for philosophers and historians of science.

Active Knowledge

I now want to introduce an action-based view of knowledge that can help us reach a better understanding of practices in science and other realms of life. But first I should acknowledge that many leading epistemologists, going at least as far back as Bertrand Russell (1912, ch. 5), have recognized that there are different kinds of knowledge, not just the propositional kind. For example, Keith Lehrer's classic text of epistemology (1990, pp. 3–4) starts by giving a list of all sorts of knowing: 'I know the way to Lugano. I know the expansion of pi to six decimal places. I know how to play the guitar. I know the city. I know John. I know about Alphonso and Elicia. I know that the neutrino has a rest mass of 0 . . .'[3] Yet, Lehrer explains that only propositional knowledge (or, knowledge 'in the information sense') is the main concern of epistemology because 'it is precisely this sense that is fundamental to human cognition and required both for theoretical speculation and practical agency'. This is an entirely typical starting point among epistemologists in the tradition of analytic philosophy. They may debate endlessly about the precise nature of justification and the exact meaning of truth, but it is rarely questioned that knowledge is a matter of *belief* in *propositions* that give *information* about the world, expressible in the form of well-formed statements.[4]

On reflection it is not clear at all that **propositional knowledge** (or **knowledge-as-information**[5]) is so much more important or fundamental than other types of knowledge that it should command the undivided attention of philosophers. I am not suggesting that the proposition-focused orthodox epistemology is *wrong*, but I do think that it is *limiting*. It obliges us to disregard many kinds of things that we readily regard as 'knowing'. One could try to argue that all these types of knowledge have less philosophical importance than propositional knowledge. As Kitcher points out, a crucial feature of propositional knowledge is that it is the most suitable form of knowledge for public communication.[6] But is that necessarily more important than the private dimension of knowing, or the tacit communication of knowledge between people in direct personal contact with one another? In the end, I am not convinced that we should expend a great deal of intellectual energy in trying to defend the primacy of propositional knowledge.

Instead, I want to endorse and develop an alternative view, which thinks about knowledge primarily in the context of action – which is to say, in the

---

[3] See also Snowdon (2004, p. 5).
[4] I will sometimes use 'statement' and 'proposition' interchangeably, except where it is important to distinguish a statement from its content, which is the proposition.
[5] I use 'information' here as a commonly understood idea, leaving it to those who see knowledge as information to settle on a more considered view of its nature.
[6] Philip Kitcher, personal communication, 29 January 2021.

context of life itself. This is a perspective articulated long ago by Marjorie Grene (1974, pp. 172, 158): 'knowing is something people do, an activity'; it is the 'full, concrete, historical person who is the essential agent of knowledge' (I will say more about the nature of epistemic agents in Section 1.2). In the context of action, the primary sense of knowing is knowing how to do something. 'I know how to do X' (distinct from 'I know how X is done' as a description not necessarily accompanied by actual ability) is a common, meaningful and important thing to say. I imagine there are roughly equivalent expressions in most human languages. This is a clear and well-established notion of 'knowing', and it should not be presumed to be reducible or subordinate to knowledge as the storage and retrieval of information. I propose to use the phrase '**active knowledge**' to designate this sense of **knowledge-as-ability**. Paying attention to active knowledge can help us greatly in making full sense of how we acquire and evaluate knowledge in science and various other walks of life.

Etymology is not philosophy, but it can give some suggestive clues: the Indo-European root *gnō* is the common source of both 'know' and 'can' in English, as well as the Scots word *ken* and the German *kennen* and *können*. The old English word *cunnan* had all of the related meanings: 'to know, know how to, be able to' (Watkins 1985, pp. 23–4; also Shipley 1984, pp. 129–33). This is in line with Wittgenstein's insight (1953, 59, §150): 'The grammar of the word "knows" is evidently closely related to that of "can", "is able to".'[7] As Anthony Kenny put it: 'to know is to have the ability to modify one's behaviour in indefinite [not pre-determined] ways relevant to the pursuit of one's goals' (Kenny 1989, pp. 108–9, quoted in Hyman 1999, p. 438).

Active knowledge is at the core of scientific knowledge. We should also note that scientific abilities are not entirely distinct or disconnected from the range of abilities that support ordinary human life, or even animal life. Active knowledge in science may simply be more systematic than active knowledge in other spheres of life, in line with Paul Hoyningen-Huene's (2013) point that systematicity is what defines science. Ordinary members of human societies know how to speak languages, make inferences, tell lies, give explanations, argue with each other, and count and sort things. Most of us know how to run, swim and kick balls. Some of us even know how to sing, paint and make pottery. We know how to recognize and remember people's faces, find our destinations through complex routes, and imagine fictional things. We know how to cook food, clean house and cut

---

[7] He adds: 'But also closely to that of "understands".'

fingernails. We may know how to grow crops, build bridges, perform surgery, raise children and teach skills to other people. Why should we deny that there is significant *knowledge* in all these activities, just because it may not exist in a propositional form?

We should want to have an epistemology that can handle such common items of knowledge, and starting with the notion that knowledge consists in *belief* will not get us there easily. We should try to deal with abilities directly, rather than skirting around them in an awkward and roundabout way, treating them as the applications of beliefs or inessential accompaniments to propositions. In the realm of purposive human action, knowers are active living and inquiring agents, and knowing is a state of such agents. Various leading philosophers have articulated the same kind of insight that I am trying to convey here, and I will try to extend and synthesize their ideas. I have already mentioned Grene, and her work developed in close connection with Michael Polanyi's (1958). Some others who have gone in similar directions are even regarded as founders of analytic philosophy – Wittgenstein, Ramsey, Ryle and also J. L. Austin. In current analytic philosophy, too, there are some congenial moves. Timothy Williamson's 'knowledge first' epistemology is right to resist a reduction of knowledge down to other notions such as belief, and to regard knowing as a mental state of the agent, namely the 'most general factive mental state'. But when he explains that being factive is a 'propositional attitude' that one has only to truths, it is clear that his focus remains on propositional knowledge (Williamson 2000, pp. 33–4). The tradition of virtue epistemology going back to Ernest Sosa is more promising, as it places the familiar account of propositional knowledge in the context of action, holding that 'judgment and knowledge itself are forms of intentional action' (Sosa 2017, p. 71). However, it seems to me that the potential of this approach is not fully realized. According to Sosa (ibid., p. 73) intentional action (an attempt) is apt if and only if it is successful because competent. This strikes me as modelled too closely on the textbook epistemology of propositional knowledge, according to which a belief is knowledge if and only if it is true because justified.

Instead of following such lines of development, I look back to the tradition of pragmatism. Pragmatist philosophers have clearly recognized the need to understand and assess knowledge in the context of action. I want to focus on what pragmatism can tell us about method-learning and practices of inquiry, pointing to a conception of active knowledge that can be useful in the philosophy of science. In Section 1.6 I will give a more detailed exposition of what I take pragmatist philosophy to be. For now, hear William James, who said long before Grene and others: 'The knower is an actor' (quoted in Putnam 1995, p. 17). John Dewey (1917, p. 12) went on to develop this vision fully, complete with his own memorable slogan: 'we live forward'. For Dewey experience is active, full of expectations and reactions, contrary to the impoverished view of it in traditional empiricism as the recording of information through sensory input. Experience is not just *given*, but *taken* by active agents. So is knowledge. Inquiry is pervasive in life, an essential activity of an organism coping in its environment, as emphasized recently by Joseph Rouse (2015).

Propositional Knowledge Embedded in Active Knowledge

In proposing to take knowledge primarily as ability, I am not suggesting that we ignore propositional knowledge. Active knowledge (knowledge-as-ability) and propositional knowledge (knowledge-as-information) are both in operation in science and everyday life. The urgent task is to clarify the relation between the two. We need to think about the *functions* of propositional knowledge, rather than just taking it as the be-all and end-all of human intellectual activity. Especially those of us who consider ourselves intellectuals should guard against taking it for granted that propositional knowledge is valuable in itself. Let's face it: possessing information is not an end in itself. Or rather, we should actually only count as 'information' what is *informative*, and what is informative depends on our purposes and our situations. In transmitting information we obviously want to achieve a high signal-to-noise ratio, but what is signal and what is noise depends on what we want to learn and why. In the words of Nicholas Maxwell, who has long been calling for a philosophy of wisdom instead of mere knowledge, 'the aim of science is not to discover truth *per se*, but rather ... *valuable* truth' (Maxwell 1984, p. 91; emphasis original). On the question of the place of human values in science there is now a robust literature in the philosophy of science.[8] This book will not be a contribution to that literature, but I do think that my focus on active knowledge can help bring the considerations of values and knowledge closer together. Thinking about knowledge-as-ability brings knowledge inescapably into the context of action, where values are immediately present. Then we can have a full view of how knowledge functions in life.

---

[8] See, for a few salient examples, Raskin and Bernstein (1987), Longino (1990), Douglas (2009), Kitcher (2011a), Carrier (2013), Brown (2020).

To refocus our view on the relation between active knowledge and propositional knowledge, let us first pull back to the general point that there are various types of knowledge, as already mentioned: knowing facts, knowing reasons (having a causal or intentional explanation for something), knowing someone or something (knowledge-by-acquaintance), knowing what it's like to be someone or to experience something (empathetic understanding), and more types besides. A common impulse concerning this situation is to understand all of these types of knowledge in propositional terms.[9] My own inclination is to recognize that various types of knowledge interact in a complex way with each other, rather than all being reducible to propositional knowledge. For example, consider what we really mean when we say we know someone. (In the sentences to follow, I propose a new gender-neutral third-person singular pronoun: e, er, em, ers.[10]) It involves the ability to recognize the person and distinguish em from other people – by er voice, by the feel of er hand, by the look of er face, by the shape of er body, and so on. It also involves knowing, implicitly and explicitly, some basic facts about em, including er habits and er typical reactions to situations. It also involves having a memory, often unarticulated, of some experiences that e and I have shared.[11]

Having recognized the complexity of the picture, I propose to simplify it in an important and useful way by conceiving active knowledge as a more encompassing category than the others (without implying a *reduction* of other types of knowledge to active knowledge). In this picture, all other sorts of knowledge depend on active knowledge, and they also contribute to it. To continue with the previous example: knowledge-by-acquaintance certainly depends on active knowledge as we have seen, and it also contributes crucially to active knowledge – how would we be able to do anything at all, if we did not know who was who and what was what?

---

[9] For sophisticated discussions in this direction, see Craig (1990), chs. 16, 17; Stanley and Williamson (2001); and Lawler (2018), ch. 2.

[10] Not only is 'e' what 'she' and 'he' have in common, but 'e' (이) just happens to be a word in Korean (my native language) indicating a 'person' gender-neutrally. It often occurs as an ending (as in 어린이 meaning 'child', and 늙은이 meaning 'old person'). I believe there is a clear disadvantage to using 'they' as both singular and plural. It is bad enough that we lost the second-person singular 'thou' long ago, later necessitating the invention of *y'all*!

[11] Knowing by acquaintance is a distinct enough kind of thing, given a different word in French and German to distinguish it from a more factual kind of knowing: *connaître*, not *savoir*, and *kennen*, not *wissen*.

From that point of view let us now return to the relation between active knowledge and propositional knowledge. It is not at all that propositional and active knowledge are opposites on the same plane, as people often seem to imply about knowledge-that vs. knowledge-how. Rather, we need to ask how propositional knowledge fits into active knowledge. The answer is twofold: propositional knowledge depends on active knowledge, and it contributes to active knowledge. Ryle gave us crucial insights on both sides of the picture. Concerning the first, he points out that knowledge-that (roughly equivalent to propositional knowledge) can be operational only when it is properly embedded in knowledge-how (similar to active knowledge). He even argues that 'knowledge-how is a concept logically prior to the concept of knowledge-that', and 'knowing-that presupposes knowing-how' (1946, pp. 4–5, 15–16).[12] In order to have any propositional knowledge one must be able to use a language. You can read this book only because you have at some point *learned to read*. The learning of one's first language requires tacit active knowledge, starting with the ability to play the pointing-game.[13] The very invention of language must have been done by people who knew how to conceptualize and communicate things in some non-verbal ways. Or consider propositional knowledge expressed in mathematical equations and formal models. What a great number of skills one needs to master in order to be able to *do* mathematics! Knowing how to multiply two large numbers is an ability, as is knowing how to solve a set of simultaneous equations, how to integrate a complicated function, or how to construct a proof. Some things in mathematics can be done by following an algorithm (if you know how to do *that*), but most mathematical work requires task-specific skills. Even if the tasks can be broken down to simpler ones, they will only bottom out in basic abilities like knowing how to count, which are not reducible to propositional knowledge. Propositional knowledge cannot stand without all sorts of active knowledge.

In the opposite direction, it is easy to recognize how propositional knowledge contributes to active knowledge. We should pay close attention to the ways in which propositional knowledge is employed by epistemic agents in their activities. Here is Ryle again (1946, p. 16): 'effective

---

[12] Alva Noë puts the point aptly (2005, p. 285): 'grasping propositions itself depends on know-how; but if know-how consists in the grasp of further propositions, then one might wonder whether one could ever grasp a proposition.'

[13] That ability is not to be taken for granted, as I have learned from eight years of trying in vain to point out to a friendly and very clever squirrel where I have put the nuts out for her. On the complexity of the pointing gesture in gorillas and humans, see Gómez (2004, pp. 186–90).

possession of a piece of knowledge-that involves knowing how to use that knowledge, when required, for the solution of other theoretical or practical problems'. That is to say, the very *point* of propositional knowledge lies within active knowledge. Ryle made a memorable distinction 'between the museum-possession and the workshop-possession of knowledge'.[14] The image of the workshop pushes us to ask what we *do* with propositional knowledge, and how beliefs (and other epistemic attitudes) concerning propositions fit into our epistemic activities, and contribute to the active knowledge embodied in those activities. Having belief in propositions is one particular aspect of knowledge, rather than its essence, or even its central core. Metaphorically speaking, propositional knowledge may only be occasional and localized crystallizations in the flow of activity that is the creation and use of active knowledge.

Operational Coherence and Its Improvement

How do we evaluate or assess the *quality* of active knowledge? This is a vexing and fascinating question, reminiscent of Robert Pirsig's (1974) struggle to understand 'quality' in *Zen and the Art of Motorcycle Maintenance*. For propositional knowledge there is one clear chief criterion by which its quality is evaluated: whether the statement in question is true. But the truth-criterion does not apply immediately to active knowledge. How, then, should we judge the epistemic quality of activities? Since active knowledge is a matter of ability, it may seem that success can serve as the main criterion for evaluation. If I claim to know how to do something, the most obvious test would be to see how well I can actually do it. But it is not quite so simple as that. I may do something successfully but only by some lucky accident, or through misconceptions that happen to work out. Not even a certified and reliable ability to do something, by itself, constitutes active knowledge. Paul Snowdon put the point memorably (2004, p. 18): 'No one would affirm that, because I can bleed or digest a three course meal, these are things I know how to do.' To say I *know* how to do something implies some sort of *understanding*, which is parallel to the element of justification that enters into the 'justified true belief' account of propositional knowledge.[15]

---

[14] This gives an overly traditional view of what happens in museums, but it was probably not far off in relation to the museums of his time.

[15] Even in relation to propositional knowledge, it should be philosophical common sense that 'knowledge goes beyond the mere possession of information' and requires a kind of understanding, a 'capacity to distinguish truth from error' (Lehrer 1990, pp. 4–5).

The understanding required for active knowledge involves a purposive aspect that is also systematic, even in relation to the simplest of acts: in order to try to make something happen, the agent has to *coordinate* carefully various movements and thoughts with each other and with external circumstances, towards the achievement of an *aim*. (This coordination may not be explicit, and the aims may not be conscious. And understanding may be something we attribute to others, as well as a matter of self-recognition.) I propose to use the term **operational coherence** to refer to such a state of **aim-oriented coordination**. Operational coherence is a key concept that I will use throughout the rest of this book, and I will spell out its meaning more fully in Section 1.4. The rough idea, metaphorically, is that a coherent activity makes sense because what goes into it all fits together nicely; the coherence consists in various aspects of the activity coming together in a harmonious way towards the achievement of its aims. In using the term 'operational' here I am giving a conscious nod to Bridgman's philosophy as noted in the Introduction. Within recent philosophical literature, the closest point of contact I have found is the work of Paul Thagard in his aptly titled book *Coherence in Thought and Action* (2000). I must stress that operational coherence is not primarily about the logical relationship between propositions. As a quality pertaining to activities, it is also not meaningful in the absence of agents who carry out purposive actions. There is a strong hermeneutic aspect to operational coherence, as it is based on how actions make *pragmatic sense* within a purposeful activity. It is a matter of *doing what makes sense to do*. Operational coherence does not reside in the 'mind-independent world', yet it expresses the empirical ('external') constraints on our thought, because the design of a coherent activity incorporates what we have learned from experience about what tends to make sense to do and what does not.

To understand the nature of active knowledge fully, we must also pay attention to the processes by which it is acquired and improved – in other words, to the epistemic activities that constitute *inquiry*. The need for attention to inquiry becomes even clearer if we ask why we want to have a theory of knowledge at all. Echoing Kitcher (2011b, p. 508), I contend that a fundamental purpose of epistemology should be to help us *get* more and better knowledge. If so, epistemology needs to tell us something instructive about the processes through which knowledge is gained, improved and evaluated, so that we can manage them better. This is the direction in which I have attempted to steer my own epistemological thinking (see, e.g., Chang 2011a).

Let's start with Charles Sanders Peirce's view that inquiry begins with a disturbed and unsettled state of doubt.[16] Dewey took this view and developed it to the full, from a more clearly action-oriented view that took inquiry as something done by an organism in its environment: '*Inquiry is the controlled or directed transformation of an indeterminate situation into one that is so determinate in its constituent distinctions and relations as to convert the elements of the original situation into a unified whole.*'[17] What did he mean by converting the elements of a situation into 'a unified whole'? I like to imagine that he was pointing to something like my notion of operational coherence, and a pragmatic sense of understanding that comes from operational coherence. Here is a very suggestive remark from Dewey (1925, p. 50): 'The striving to make stability of meaning prevail over the instability of events is the main task of intelligent human effort.'[18] So some sort of *sense-making* is clearly seen by Dewey as an important aim of inquiry, to overcome the initial puzzlement that sets inquiry off.

Inquiry is a striving towards greater operational coherence. In order to escape the state of disorder that prompts inquiry, we must change something about the situation in order to create more operational coherence in the activities that we can perform in the situation. This adaptation takes the form of **aim-oriented adjustment**, a concept that I will explain fully in Section 1.5. This adaptive process is driven at each moment by the relief and satisfaction provided by increasing coherence, without a fixed final destination. Anything and everything that is within our power to change may be changed in the process of increasing operational coherence. There is nothing in our knowledge that is fixed and validated forever and unconditionally, and inquiry does not follow pre-determined and eternal methods. Inquiry is a process of method-learning as well as content-learning. We are truly floating in Neurath's boat, 'like sailors who have to rebuild their ship on the open sea, without ever being able to dismantle it in dry-dock and reconstruct it from the best components' (Neurath [1932/3] 1983, p. 92).

---

[16] Peirce (1877, p. 5). See also Peirce (1986), W3:248. According to Cheryl Misak (2013, pp. 32–3), 'in Peirce's view, what is wrong with the state of doubt is . . . that it leads to a paralysis of action'.

[17] Dewey (1938, pp. 104–5); emphasis original. Rouse (2015) is an exemplary inheritor of this view of inquiry among our contemporary philosophers. Taking inquiry explicitly as a process, Dewey laid out the following steps of it (1938, pp. 105–12): antecedent conditions; institution of a problem; the determination of a solution; reasoning and the examination of meaning; and the development of an idea 'in terms of the constellation of meanings to which it belongs' (ibid., p. 112).

[18] I thank Céline Henne for pointing out this passage to me, and for numerous other pointers and insights concerning Dewey's work.

All attempts to locate certainty at the foundation of our knowledge having failed, various philosophers in the early twentieth century articulated the realization that the traditional ideal of certainty must be removed if we are to forge a realistic ideal of knowledge. These included Wittgenstein, especially in the notes published posthumously as *On Certainty* (1969), and Dewey in *The Quest for Certainty* (1929). If the struggle to settle an unsettled situation is sufficiently successful, it will result in a situation that is sufficiently settled for helping us launch further inquiries that are more restricted. Such restricted forms of inquiry, including fact-gathering and learning how to perform certain well-defined tasks, are the types of work that Kuhn put under the rubric of 'normal science', which can happen once a paradigm is well-entrenched. These may be the more readily recognized sort of inquiry, but they can only function after some successful unrestricted inquiry has been carried out in order to fix the framework within which they are carried out, as Céline Henne (2022) discusses in making her distinction between 'framing' and 'framed' inquiry.

Plan of the Rest of the Book

What next? I hope you will want to read on, and there are different options for how to proceed. If you want to get a full and detailed treatment of the themes discussed so far, please continue with the remaining sections in this chapter (or at least some selected sections), which will discuss various specific aspects of active knowledge in more depth and detail: epistemic agents (Section 1.2), epistemic activities (Section 1.3), operational coherence (Section 1.4), and inquiry (Section 1.5). I will also explain the affinity of my ideas to pragmatism (Section 1.6).

If you are already quite persuaded by the programme laid out so far and anxious to find out the rest of the story, you might want to go directly on to Chapters 3, 4 and 5. Just take in the overview (Section 1) of each chapter if you have limited time and interest, but I hope you will get into some of the further sections as well. Selective reading can be guided by the summary at the beginning of each section. (In order to distinguish visually the friendlier surface-level summaries from the more specialist and detailed discussions, I have put the Introduction, the overview of each chapter, and the summary of all the other sections in a different typeface.) Chapters 3 and 4 will show how I use the notion of operational coherence in order to reconceive the very notions of reality and truth. Building on these views of truth and reality, Chapter 5 will present a doctrine of realism that is suitable for realistic people.

But before I get to all of that, a *clearing* needs to be made, where these ideas can have space to grow. This will be the task of Chapter 2, which will try to show how we can and should get away from the well-entrenched notion that our knowledge of nature is a matter of *correspondence* between our theories and the ultimate reality 'out there'. That will be the only primarily negative and critical chapter in the book, the other chapters being mainly devoted to the positive exposition of my own ideas. If you feel that the unorthodox approach to knowledge that I have laid out so far is unnecessary because more orthodox approaches will do the job equally well or better, please do read Chapter 2 before other chapters (or before abandoning the book altogether). On the other hand, if you are in agreement with my general approach and would like to see how I defend it against orthodox views, you may also find Chapter 2 interesting and instructive for that purpose.

1.2 Epistemic Agents

To understand the nature of active knowledge in a philosophically rigorous way, we need to have a clear and precise characterization of epistemic activities. That task begins with understanding knowers as full-fledged agents. Epistemic agents do not simply possess beliefs and desires, which are the chief notions employed in mainstream discussions in the philosophy of action. They are beings also endowed with certain physical and mental capacities, who engage in purposive actions, and make genuine choices and judgements. Epistemic agents are embedded in social communities that embody and enforce certain normative standards. There is an iterative process of emergence through which individuals arise from society and society is constituted through interactions between individuals, ascending to ever-higher levels of sophistication.

Basic Properties of Epistemic Agents

A full account of active knowledge requires a good *ontology* of epistemology. There is a tendency in the philosophy of science to present the scientist as a ghostly being that just has degrees of belief in various descriptive statements, which are adjusted according to some rules of rational thinking (e.g., Bayes's theorem) that remove any need for real judgement. Whatever does not fit into this bizarre and impoverished picture, we tend to denigrate as matters of 'mere' psychology or sociology. We need a more serious understanding of scientists as agents, not as passive receivers of information or algorithmic processors of propositions. With a brief sketch of epistemic agents here, I want to at least express a recognition of the issues that we need to think about.

Start again with the motto from James and Grene: the knower is an actor. The first step is to recognize purposive behaviour in the knower, at least in terms of instrumental rationality. The most basic thing about an agent is that e acts purposefully (or can be understood as acting purposefully), striving to achieve certain aims that are formulated on the basis of er desires and beliefs. The agent takes the kind of actions that e believes will contribute towards the satisfaction of er desires. That is the 'standard story of action' in philosophy, as Jennifer Hornsby calls it, according to which actions are 'belief-desire caused bodily movements' (2007, p. 180, also p. 165). Hornsby (2004, p. 2) has criticized this account strongly as not giving a truly active role to the agent, 'not a story of agency at all'. But actually, in philosophy of science even just a proper recognition that the epistemic agent has desires, rather than just beliefs, would be a significant advance. There are many types of pleasure that motivate scientists (and other human beings), including physical comfort and sensual well-being, abstract and concrete understanding, love and conviviality, self-esteem, security, legacy, and a sense of beauty, order and coherence. And on top of that, of course, we have to think about the things that various people have regarded as 'the aim(s) of science': truth, empirical adequacy, economy of thought, etc., which are best understood in terms of values. In order to understand concrete practices, we need to see how such desires and values shape the specific proximate aims that drive particular epistemic activities, as I will discuss further in Section 1.3.

Aside from desires and values, epistemic agents have beliefs indeed. But we need to think about much more than just explicit and articulated beliefs. There are also things we take for granted without examination, and such presumptions are necessary for enabling any kind of action. Of particular importance are expectations concerning the future. Expectations are often not beliefs at all, if by belief we mean a conscious assent to an articulated proposition. Expectations often exist on the 'horizon' as Edmund Husserl would have it (see Føllesdal 1990), or, in the tacit dimension in Polanyi's way of thinking. They can even consist in *not entertaining* certain possibilities. When I am walking along as normal, my expectations involved in that activity will not be exposed or even formulated until they are met with something incoherent with them, such as the tremors of an earthquake, or the left arm grabbed by an excited old friend, or a gaping hole in the pavement. Scientific practice is also full of expectations, sometimes guiding our activities smoothly, sometimes preventing certain activities, sometimes making us attempt something repeatedly without a clear sense of why.

Something often neglected in philosophical accounts of science is epistemic agents' capabilities, or capacities, which are considered mostly in the discussions of ethics and human flourishing. In a related way, capacities are discussed in terms of legal responsibilities, or cognitive capacities in the philosophy of mind. Hornsby criticizes the standard story in the philosophy of action for leaving out agents' capacities, too:

> When abilities are allowed a place in the explanation of action, it becomes clear how narrowly focused are the explanations from agents' reasons given in the standard story. I said that it would be a sort of magic if someone's intentionally doing something were consequential merely on their having a desire and a belief. (Hornsby 2007, p. 170; see also 2004, pp. 20–2)

It is important to recognize both mental and physical capabilities here, and also their mutual entwinement, remembering Polanyi's (1958, ch. 4) emphasis on the role of skills in scientific work, with due attention to embodiment. Most of the specific capabilities have to be learned, so we need to consider the process of learning and training. The consideration of capabilities needs to enter the discussion of a wide range of issues in the philosophy of science, such as observability, testability, simplicity and incommensurability – and therefore also realism, demarcation, confirmation, theory-choice and so on. Considerations of scientific rationality are greatly hampered if we do not consider what the capabilities of the agents involved are, because the judgement of what is rational for them to attempt depends greatly on what they are in fact able to do (remember the old lesson in ethics: 'ought implies can').

The last aspect I want to stress in the ontology of epistemic agents is the fact that they make choices and judgements. The standard story of action does not seem to leave any room for judgement, regardless of the types of cause that the explanations appeal to: utility-seeking, cognitive-psychological, neurological, sociological, what have you. For example, in an interest-based sociological explanatory schema, the picture of the individual is just as impoverished as that of the utility-maximizer in the individualist rational-choice theories. Instead, we need to find ways of giving some substantive meaning to words like 'choice' in the phrase 'theory choice', and 'decision' in 'decision theory' (cf. Bradley 2017). Even if the correct account of actions would be ultimately deterministic, decisions to act in specific ways are determined by a distinct combination of aims, beliefs and capabilities for each individual and each community. This will create at least an appearance of freedom on the part of the agents, and make the reality of scientific practice pluralistic.

The Social Dimension: Iterative Emergence

It is crucial to take into account the social nature of epistemic agents. This is no place to attempt a full social ontology, social epistemology or sociology of knowledge, but I do want to offer some pertinent reflections that I have found useful in framing the discussions to follow in the rest of this book. It is an urgent future task to improve my idiosyncratic home-made social theory with better attention to the considerable existing literature (for a notable recent synthesis see Epstein 2015). As Helen Longino (forthcoming) stresses, it is crucial for social epistemology to pay attention to interactive processes, and individuals only become epistemic agents through deliberative interactions.

Here is an unlikely source of inspiration. In the chapter on 'conviviality' in his classic *Personal Knowledge*, Polanyi laid out the necessity of the social in the epistemic. He stressed the crucial role of the 'civic coefficients of our intellectual passions' (1958, p. 203), such as the 'sharing of convictions', the 'sharing of a fellowship', 'co-operation' and the 'exercise of authority or coercion' (ibid., p. 212). These operate at a tacit level in the first instance, effective even in various non-human animals. This is an often neglected aspect of Polanyi's thinking: the tacit is embodied and individual, but also inherently social.[19] Wittgenstein and Polanyi concurred that knowledge must be founded on the trust we place in others, and the store of facts we rely on necessarily rests on the testimony of others (see Daly 1968). Martin Kusch (2002) and others have built on this tradition greatly.

The recognition of the necessity and priority of the social should not be flattened into the thought that 'everything is social'. For something like scientific knowledge to arise, we must have independent individuals as well as unindividuated collectives. The society–individual interaction needs to be conceived in an iterative way, avoiding reductionism in both directions. Individual action and cognition are grounded in society, but they are not *merely* social, and we should not presume that they are explainable or even fully describable by means of collective factors alone. The rational individual agent arises from the social matrix, but with a capacity for independent thought and dissent. And from the association of such individuals emerges a higher level of sociality that forms an integral aspect of life as we know it, and also forms the basis of advanced systems of knowledge. This iterative emergence of the individual from the social and the social from the individual continues indefinitely.

---

[19] And there are remarkable abilities that non-individualist and mostly inarticulate society can achieve, as in the case of bees and ants.

The first thing to understand in this whole picture is the process of individuation.[20] How does the distinctive individual come to exist? Each developing individual has a particular physical and mental make-up that is different from others, and a life trajectory that is inevitably different from anyone else's. The individual also has a capacity for seeking operational coherence, for *pragmatic sense-making* that harmonizes er thoughts and actions with er make-up and circumstances. What makes sense for each person is different, and this prompts *dissent*, an inner voice that says 'no' to what others say, an inner revulsion against social expectations. This is as real as the submission-instinct shown in the Milgram experiment that Barry Barnes (2000) makes so much of. If the social is entirely satisfactory, there is no need for the individual. The individual defines erself *against* the social.

Individuation is especially important when we consider science. I would even say that the inquiring attitude is essentially linked to the emergence of the individual. In an important sense, science begins with dissent and critique. This is tied with Dewey's (1938) notion that inquiry begins with an unsettled situation. Inquiry wouldn't start at all without being motivated by some individual's dissatisfaction. There is also a deep connection here with the fundamental empiricism in science, which is founded on the authority of observation, which has an irreducibly individual, even private, dimension. Saying 'let me check for myself' is the beginning of the scientific attitude, and it is a fundamentally *individualist* declaration (Bridgman 1940; Pritchard 2016). This is not the old methodological individualism, which ignores the social grounding of individuals. Rather, it is an activist individualism promoting and celebrating the rise of the autonomous individual standing up on the social ground.

But this is certainly not the end of the story. From the association of self-actualizing individuals emerges a higher level of sociality. In this process there is one aspect that deserves particular attention: *second-person interactions*, in which I treat a fellow member of society as an individual like myself. From such interactions arises something truly irreducible to individuals. The importance of the second-person standpoint in ethical life has been emphasized by Stephen Darwall (2006), and of course long ago by Martin Buber (1923), but it is also crucial in epistemology and the philosophy of science. There is a deep presumption underlying any second-person interaction: you are an individual, with the basic rationality that consists in the

---

[20] This is a term I once absorbed from Jungian psychology, but it does not need to be placed in that precise context. It is related to what other psychologists have called self-actualization or ego development.

coherence of belief and action, and the basic cognitive capacity required for communication, also with at least a minimum degree of good will and conviviality. We often lose sight of second-person interactions in the usual clash between individualistic and social perspectives, but so many of the common speech acts in life are of the second-person variety: commands, questions-and-answers, assertions-and-(dis)agreements, arguments, and explanations. Many key epistemic activities, too, are inherently in the second person. Describing or explaining is meaningless and pointless unless there is at least some imagined 'you' to whom it is directed. In philosophy and science we talk too often about questions and answers without thinking about the second-person interaction of asking-and-answering. Even more problematic is the divorcing of 'arguments' from *arguing*, which is something you and I engage in, and the removal of 'justification' from the persuasion of me by you. It will not do to reduce all of these things to the flatland of propositions and their logical relations.

A social group develops a thicket of second-person interactions. What Wilfrid Sellars called 'we-intentions', adapted later in different ways by Raimo Tuomela and John Searle, must emerge from you-and-I interactions.[21] In this thicket of interactions people establish customs, routines and institutions. We now have a higher level of sociality, composed of individuals who consciously connect with each other in order to live better together. Society, so formed, shapes individuals in turn, in a process similar to the earlier-stage shaping of individuals through inarticulate socialization. When patterns of social interactions become settled, the social norms, routines and expectations may become so internalized as to no longer require explicit reinforcement. If so, the individual becomes submerged into society again – as a willing and comfortable master of unspoken rules, or a timid conformist, or a half-comprehending misfit. What was once consciously negotiated becomes *sedimented* (Husserl [1954] 1970; Føllesdal 2010) – that is, added to the stock of shared unarticulated culture. Such a process happens in science, too: think of people who use thermometers, clocks or pH meters with no thought as to how the standards are established and maintained.

This formation of higher-level society by conscious agreement is still not the end of the story. A higher level of individual struggles to emerge out of that newly sedimented sociality. This cycle of socialization and individuation can continue indefinitely as societies and individuals become ever more complex and sophisticated. And a backwards glance also shows that the initial

---

[21] See Schweikard and Schmid (2021), esp. sec. 2.3, on Sellars's and other related ideas.

social dimension with which I started my discussion was not the very beginning of the process. There were individual animals before then, each of which was shaped in a proto-social environment; and so on.

1.3 Epistemic Activities and Systems of Practice

To facilitate the philosophical understanding of active knowledge, we need to craft good *units of analysis* that can accommodate aspects of knowledge and inquiry that are not captured by propositions. For this purpose I propose the notion of **epistemic activity**. An activity is not a one-off act, but a programme of action designed for the achievement of a recognizable aim. An activity has an inherent aim that partly defines it, and also various external functions. Various activities can be pulled together in order to form an integrated activity, with a new overall aim. However, the relationship in such integration is not reductive, and the contributing activities are not necessarily simpler than the overall integrated activity. When we are considering extensive and complex fields of work such as science, the most important unit of analysis is a **system of practice**, which is a coherent network of activities comparable to a Kuhnian paradigm. A system of practice, unlike an integrated activity, has multiple aims. Activities and systems both exhibit operational coherence, a concept that will be characterized in more detail in Section 1.4.

Activities and Practices

The usual unit of analysis in epistemology is a proposition, or a theory conceived as a collection of propositions, but these are not appropriate for active knowledge. In order to make full sense of the notion of knowledge as ability, we need suitable notions of action and practice. What exactly is doing, and what are those somethings that we do? Before I try to explain my own concepts of epistemic activities and systems of practice, I should comment briefly on some relevant bodies of work in the philosophy of action, and in theories of practice in the social sciences and in science and technology studies.

Within the literature in the philosophy of action I look particularly to the work of Jennifer Hornsby, who emphasizes two points that are fundamental to my thinking. First, she stresses the importance of attending to *activities* (Hornsby 2007a, p. 170): 'Human agents participate not only in once-off [sic] actions, but also in activities.' An activity is a complex, organized and regulated series of doings. While an individual act may be performed in a haphazard way, to call something an 'activity' implies routinized and repeated doings directed at an aim, following a reasonably stable set of rules and norms. Second, Hornsby focuses on capacities and on knowledge-as-ability: 'a person's knowledge of how to do things informs more than token actions', and 'much human agency is made possible by people's possession of capacities which ensure that they have standing abilities to engage in one or another activity' (ibid., p. 179).

There is a large and diverse body of literature theorizing about the nature of practices.[22] One question to be addressed before I go on to elaborate my notions of 'epistemic activity' and 'system of practice' is whether I shouldn't just speak about practices, which would also help connect my thoughts more straightforwardly to existing discourses. One difficulty is that people have meant all sorts of different things by 'practice', so it is impossible to be precise about it without further specification. The best attempt at a concise and fair summing-up that I have found is by Theodore Schatzki (2001a, p. 11), who states that practices are commonly conceived as 'arrays of activity', or more specifically, as 'embodied, materially mediated arrays of human activity centrally organized around shared practical understanding'. In that sense a practice is akin to what I call a system of practice, but in other uses a practice designates something more akin to an epistemic activity.

Rouse (2001, p. 199) makes an important distinction between mere regularities in people's behaviours and normative structures that govern people's behaviours. Only the latter deserve to be called 'practices': 'actors share a practice if their actions are appropriately regarded as answerable to norms of correct or incorrect practice'. I follow Rouse in adopting the normative view of practice. According to Schatzki (2001b, pp. 60–1), the normativity of practice is a 'teleoaffective structure', which combines a specification of ends and values. I also follow Rouse, Schatzki and others in emphasizing the place of understanding in practices, of which I will speak further in Section 1.4. Another important dimension of practice to note, especially for the purpose of philosophy of science, is the fact that most of our practices involve engagement with objects other than ourselves. As Rouse (2001) and Karin Knorr Cetina (2001) both stress, this objectual engagement brings open-endedness, contingency

---

[22] As my principal guides I follow two collections: *The Practice Turn in Contemporary Theory* (Schatzki et al. 2001), and *Science after the Practice Turn in the Philosophy, History, and Social Studies of Science* (Soler et al. 2014). The editors' introduction to the latter volume provides a masterful and comprehensive survey of the history and the main characteristics of the 'practice turn' especially as it pertains to science studies. Also very instructive are the accounts by Michael Thompson (2008) and David G. Stern (2003). Within the philosophy of science, a very important yet sadly neglected work in this direction is Harré and Llored (2019).

and creativity to practices. All of these are lessons that inform my conception of epistemic activities and systems of practice.

Epistemic Activities

In my work in the history and philosophy of science I have been trying to put into practice the 'activity-based analysis' of scientific knowledge (Chang 2011a; 2014). For the sake of consistency and continuity, I will begin by quoting from my best-publicized previous attempt to characterize epistemic activities, from my book *Is Water H₂O?* (Chang 2012a, pp. 15–16):

> An *epistemic activity* is a more-or-less coherent set[23] of mental or physical operations that are intended to contribute to the production or improvement of knowledge in a particular way, in accordance with some discernible rules (though the rules may be unarticulated). An important part of my proposal is to keep in mind the aims that scientists are trying to achieve in each situation. The presence of an identifiable aim (even if not articulated explicitly by the actors themselves) is what distinguishes activities from mere physical happenings involving human bodies ...

There I noted a similarity with Soler's notion of an 'argumentative module', which is 'individuated and defined as a unit on the basis of its aim: on the basis of the question it is intended to answer, or the problem it tries to solve' (Soler 2012, p. 235).

I will now develop the concept of epistemic activity further, starting with some simple modifications. The notion of 'epistemic' should be broadened to include the evaluation and use of knowledge, as well as its production and improvement. And it should be stressed that most of the time mental and physical activities are combined together. In addition to mental and physical activities, Bridgman (1959, p. 3) also noted the importance of 'paper-and-pencil operations' in science; similarly, Ursula Klein (2003) has stressed the importance of the use of 'paper tools' in chemistry.

One major issue to be worked out is the different levels of aims. Specific activities are designed to achieve proximate aims that help agents achieve their ultimate desiderata. Suppose someone asks me, while I'm striking a match, what I am trying to achieve. My answer may be 'I'm trying to light a match', or 'I'm trying to get a combustion-analysis of an organic compound

---

[23] In a slightly earlier formulation (Chang and Fisher 2011, p. 361) I used the word 'system' instead of 'set'; that usage is avoided here, as it is not consistent with the explication of 'system of practice' that I have now made.

going.' Both are cogent answers, but they get at two different kinds of aim. The first answer addresses what I will call the **inherent aim** of the activity: getting the match to light is the whole point of the activity itself, regardless of why one is engaged in that activity – that may be in order to light a Bunsen burner with it, or to burn down a house, or just to watch and admire the marvellous phenomenon that combustion is. These latter reasons might be called the **external functions** of the activity, as they are consequences of the successful execution of that activity. An activity is *defined* partly in terms of its inherent aim, which exists regardless of any external functions that the activity may serve (match-lighting is not match-lighting if one does not at least intend to light a match).

Finally, I should stress that any description of an activity we can give is a *programme* of action, whether in terms of retrospective understanding or as prescriptive guidance. Such a programme is bound to be abstract, in the sense of not including all the features that are present in each instance of its execution. So, any activity that we can describe is not precisely instantiated in our actual doings, and there is no uniquely right way to identify and classify activities out of the stream of doings that we continually carry out in life. In this sense, epistemic activities (and also systems of practice) are 'ideal types' in Max Weber's sense of the term. An ideal type is a concept derived from observable reality but with conscious simplification and exaggeration. Weber says: 'an ideal type is formed by the one-sided accentuation of one or more points of view' according to which 'concrete individual phenomena ... are arranged into a unified analytical construct' (quoted in Kim 2019). As Sung Ho Kim explains (2019, sec. 5.2): 'Keenly aware of its fictional nature, the ideal type never seeks to claim its validity in terms of a reproduction of or a correspondence with reality. Its validity can be ascertained only in terms of adequacy.' As I will discuss further in Section 1.4, there is a hermeneutic dimension to activities: the operational coherence of an activity is about pragmatic sense-making, on the part of either the agents themselves or others who analyse their actions.

Systems of Practice

Epistemic activities normally do not, and should not, occur in isolation. Often they form a network that is dense enough and large enough to deserve to be called a 'system of practice'. To refer back to my 2012 publication again:

> A *system of practice* is formed by a coherent set of epistemic activities performed with a view to achieve certain aims ... Similarly as with the

coherence of each activity, it is the overall aims of a system of practice that define what it means for the system to be coherent. (Chang 2012a, p. 16; also Chang 2014, p. 72).

I now want to consider more carefully how activities come together coherently to constitute a system of practice. (Any kinds of activity may come together to form a system of practice; I tend to speak of epistemic activities since I am mostly addressing issues relating to knowledge.) The various activities in a system of practice are not merely performed in the same general setting; rather, they come together in very particular ways to meet certain aims. The system coordinates various activities for the satisfaction of the system-level aims.

Unlike an activity, a system of practice does not have one single inherent aim. In fact, that may be considered the chief difference between an activity and a system. Recall the example introduced in Section 1.1: the chief overall aims of the Lavoisierian system of chemistry were the knowledge of the composition of various substances, a good classification of all chemical substances, and the explanation of chemical reactions. There is both dependence and independence among such aims. Lavoisier's was a 'compositionist' system of chemistry (see Chang 2011b; 2012a, sec. 1.2.3): knowing the compositions was essential for the other two aims of explanation and classification, because in this system there was a commitment to make explanations and classifications on the basis of compositions. On the other hand, classification and explanation were largely independent aims, in the sense that one could be pursued without the other at least to an extent.

To take a different kind of example: consider the game of soccer – not an individual match, but the 'game' as an institution. This may be considered a system of practice. The whole institution of soccer does not have a unitary inherent aim, so it is not a single activity, while particular activities *within* soccer have unitary inherent aims: for example, the inherent aim of goal-keeping is to prevent the other team from scoring, and the inherent aim of passing is to give the ball to another player on one's own team. But isn't *winning* the inherent aim of the game itself? That may be said to be the aim of each team's engagement in a match, but not of the whole sport, to which 'winning' does not apply. If we ask seriously about the aims of the whole system of soccer, there is no one clear answer, but multiple answers: to provide entertainment for the people, to make income and profits for some individuals and entities, to promote health and fitness in society, to contribute to community solidarity, etc.

The Coordination of Activities

Now, there is something similar between how different activities come together to serve system-wide aims, and how an activity is formed by the coordination of many different doings. More thought is needed on the coming-together of doings. When I previously characterized an activity as 'a set of operations', I did not define what an 'operation' was, thinking that it was innocuous enough as something similar to 'activity' but as a convenient term for sub-components of an activity (Chang 2014, pp. 74–5). More problematically, I implied an overly atomistic picture, as pointed out in Léna Soler and Régis Catinaud's (2014, p. 82) cogent critique of my ideas. I believe that I have now found a way of avoiding the atomistic perspective.

A central puzzle is that even the simplest activities seem complex and susceptible to further analysis in terms of constituent activities, with seemingly no end to such analysis. For example, take one activity from the Lavoisierian system of chemistry: combustion-analysis. This activity incorporates various other activities: setting something on fire, capturing the combustion-products (gases) by means of other chemicals, weighing the resulting compounds, and making percentage-calculations. But even those activities in themselves seem to consist of other activities. For example, the activity of weighing-with-a-balance includes the placing of samples and weights on balance-pans, reading the number off the scale, and certifying the weights used as correct standard weights. But the certification-activity, without which the whole activity of weighing lacks validity, is itself a very complex thing! It may consist in ordering the weights from a reliable supplier, or comparing them to a more trusted set of weights, or checking them against certain natural phenomena (e.g., the weight of a certain volume of water at a certain temperature, in the conception of the originators of the metric system). Whichever method we opt for, it seems clear that this 'component' activity of weight-certification is not going to be simpler than the whole activity of weighing-with-a-balance.

The analysis of activities is not an atomistic or reductionist enterprise. It isn't quite right to say that an activity is 'composed of' simpler activities, since 'composition' implies an atomistic ontology too strongly. I prefer to speak of *constituent* activities, rather than *component* or *elementary* or *basic* activities. There is no lowest level of description, no rock-bottom of atomic activities, and no clear end to the process of analysis. Yet in many situations we do gain useful insights from analysing an activity into its *apparent* constituents, and the analysis should be carried out wherever it is productive, and only and as far as it is productive.

So how is the incorporation of activities to form another activity done? A couple of social metaphors might be instructive here. The United Nations isn't simply all of its member nations put together, because the functioning of the UN requires shared institutions, routines and purposes, which do not exist within individual member nations. Nor is it the case that the member nations wholly belong to the United Nations. So the whole here is both more *and* less than the sum of the parts. Likewise, when individuals come together to form an association (Neighbourhood Watch, Alcoholics Anonymous, or what have you), the association is not just the individuals put into a set, and each individual member has a very complex life outside the realm of the association, and e also belongs to a number of other associations. It is similar when activities come together to form another activity. In order to bring the constituent activities together, something outside those activities must be imposed in order to connect them; the most important factor is the overall aim of the integrated activity, which the constituent activities are brought together to serve.

Now I can attempt to present an overview of the ontology of epistemic activities and systems of practice. An activity is identified and individuated by its operational coherence in relation to a unitary inherent aim (and may have various external functions). Typically, an activity can be seen to incorporate other activities into it, to serve its own overall aim while the constituent activities each retain their own inherent aims, too. The constituent activities, in turn, incorporate other activities. We may pursue the analysis of activities into their constituents indefinitely, but this is not a situation of infinite regress, because constitution here is not reductive. Rather, the overall picture is more of a reticular one, a network of a great number of activities, each one incorporating various others, reminiscent of the 'rhizome' structure that Gilles Deleuze and Félix Guattari ([1980] 1987) speak about. As for a system of practice, it is formed by the coordination of activities without a single overarching aim – if there were a unitary inherent aim, then the coordinated whole would constitute a single integrated activity. A system of practice has multiple system-level aims, and each of those aims is not locatable within any one of the activities involved. The coherence of a system of practice consists in an effective coordination of the external functions of various activities for the achievement of system-level aims. How well the different aims within a system go together is also a matter of operational coherence.

1.4   Operational Coherence

Operational coherence is a key notion in my account of active knowledge, and a core part of the definitions of epistemic activity and system of practice, as seen in Section 1.3. Now I will try to give a more precise and detailed characterization of this concept, especially because it will play a crucial role in the reconceptualization of the notions of reality and truth in Chapters 3 and 4. In previous publications I defined operational coherence in terms of the harmonious relationship among the operations constituting an activity. By grappling with some puzzles arising from that definition, I now arrive at a more fundamental view. In short, operational coherence consists in **aim-oriented coordination**. A coherent activity is one that is well designed for the achievement of its aim, even though it cannot be expected to be successful in each and every instance. Operational coherence is based on pragmatic understanding; it consists in doing what *makes sense* to do in specific situations of purposive action.

What Is Operational Coherence? Intuitions and Illustrations

In the overview of this chapter (Section 1.1) I put forward operational coherence as a chief criterion for assessing the quality of active knowledge. In Section 1.3, I presented it as a key characteristic of epistemic activities. I will now give a more detailed and considered characterization of operational coherence. Let me start by motivating the concept again with some illustrative examples. Operational coherence is a pertinent concept in all sorts of activities, scientific and quotidian. In daily life we employ literally thousands of simple skills that require a good coordination of bodily movements, material conditions and mental concepts: drinking a glass of water, tying shoelaces, eating with chopsticks, riding a bicycle – or walking up the stairs, which is quite an achievement as contemporary robotics has learned. Take a very simple activity like match-lighting (which was, incidentally, so essential to the progress of chemistry, and even physics, for so long!). Most people can probably bring up the memory of learning how to light a match, which actually takes a surprising degree of skill and coordination to do well. With one hand I hold the matchbox steady and firm, with the rough strip facing my other hand, in which I hold the match tightly, just so; I pull the head of the match across the rough strip on the box (and I break the matchstick – no, no, the correct move is to push it), at an appropriate angle and at the right speed with some abruptness; I stop the movement of that hand promptly once the flame comes on. I need to bring these operations together well enough for my match-lighting activity to be coherent. It is important to keep in mind that coherence doesn't pertain to a single act, but to a sustained and organized *activity* (and even a whole system of practice).

The same kind of coordination takes place in scientific and technological practice, only with more theorized, complicated and careful actions. An extreme case is the operation of the global positioning system (GPS), as discussed by Peter Galison (2003, pp. 285–9). GPS requires an intricate coordination of a range of material technologies (geostatic satellites, atomic clocks, electromagnetic signals, mobile phones, etc.) and various abstract theories (Newtonian mechanics to fly the satellites, quantum mechanics to run atomic clocks, and both special and general relativity to make corrections to the atomic-clock readings depending on the speed of the satellites and their locations in the earth's gravitational field). Each element of this enormously complicated set-up is carefully coordinated with the other elements, to enable a marvellous degree of operational coherence in the activities we undertake by means of it. And even in this theory-heavy set-up, it is not the case that the operational coherence follows from a single unified theory (most fundamentally, we have no unified theory of quantum gravity, not to mention a theory that encompasses both the theoretical physics and the engineering systems involved in GPS). Rather, the coherence is achieved in a highly ad hoc manner, applying selected aspects of various theories to different parts of the system in a judicious way designed to achieve the specific aims at hand.

In puzzling out what operational coherence is, it is also helpful to think about what happens when coherence is lacking. If I try to drink water by directing the glass to my nose, that is an incoherent activity. When we do not heed the sign that warns 'Mind your step', that rare moment of stumble reminds us how well we normally maintain the coherence of our bodily movements and adapt them to our external surroundings without even thinking about it. Another example, from the typical social life of a professional (in the days before the Covid-19 pandemic): you go to a conference, meet a colleague that you really like but don't know very well; you offer a warm handshake, your colleague offers a discrete hug or kiss; your greeting ends in an incoherent tangle. Incoherence may sometimes be traceable to false or mutually contradictory beliefs, but ineptitude of belief is certainly not the only reason for it. Incoherence can also arise due to the lack of basic capability (starting with weak eyesight or muscular strength), the use of inappropriate materials, poor timing between different operations, the application of mutually conflicting rules, and so on. And the examples just cited should make it quite plain that operational coherence is a matter of degrees, and not a precisely quantifiable one at that. It is necessarily a less precise concept than consistency, which is well defined through logical axioms.

Before going on further, I want to anticipate a common worry: the so-called coherence theories of truth and justification in epistemology may slide into relativism, idealism or constructivism. If coherence is simply a matter of a positive mutual relationship between our beliefs without anything else to ground any of them, then there is a legitimate concern that coherence does not provide any link between knowledge and reality. In the most simple-minded version of the coherence theory of truth, coherence is taken to mean mere logical consistency within a set of statements. James O. Young (2015) notes that more plausible versions take the coherence relation as 'some form of entailment' or 'mutual explanatory support between propositions'. Similarly Richard Foley (1998, p. 157) says, in relation to justification: 'Coherentists deny that any beliefs are self-justifying and propose instead that beliefs are justified in so far as they belong to a system of beliefs that are mutually supportive.' Catherine Elgin (2017, pp. 71–3) also has a liberalized notion of coherence, laying out the minimum requirement that 'the components of a coherent account must be mutually consistent, cotenable, and supportive', and also considering the relations between multiple levels of commitment. However, even in the more sophisticated versions of coherentism, the problem of circularity remains.[24] *Operational* coherence is a wholly different matter. It cannot be achieved arbitrarily by decree, wishful thinking, or mere agreement among beliefs or people. On the contrary, in order to do things coherently we need to have an understanding and mastery of our surroundings. Operational coherence carries within it the constraint by nature. Through operational coherence the world outside the control of the mind is brought to bear on knowledge. In fact, in Chapter 2 I will argue that operational coherence is the only means by which reality can shape our practices, and in Chapters 3 and 4 I will show how operational coherence can ground the very notions of reality and truth.

Three Puzzles

In an earlier publication I defined 'operational coherence' as follows:

> an activity is operationally coherent if and only if there is a harmonious relation-
> ship among the operations that constitute the activity; the concrete realization of

---

[24] Note, however, Thagard's point (2000, p. 77) that circularity is not necessarily regress: 'Coherence-based inference involves no regress because it proceeds not in steps but rather by simultaneous evaluation of multiple elements.'

a coherent activity is successful, *ceteris paribus*; the latter condition serves as an indirect criterion for the judgement of coherence. (Chang 2017b, p. 111)

This definition of operational coherence is not particularly wrong, but it needs further development and reorientation. Since the old definition is out there in print, I feel obliged to show how my current ideas evolved from there, rather than just presenting the formulation that I have now reached. And aside from doing penance for having published a half-baked definition, there may actually be some benefit in showing how the recent changes that I have made were motivated.

My old definition of coherence left three puzzles unsolved. First, there is a problem with conceiving operational coherence as a relationship among the operations that constitute an activity. As Soler (2009, ch. 9; 2012) has convinced me, it is unhelpfully constraining not to allow ourselves to think of coherence in terms of the harmonious relationship between different *types* of aspects or elements of an activity (such as theoretical assumptions, bodily abilities, perceptions, social constraints, and properties of our tools). Still, it is difficult to think cogently about how such a heterogeneous set of things relate to one another, which is what had originally pushed me towards the onto- logical homogeneity of dealing only with operations. But that solution only masked the difficulty in any case, because making sense of the interaction between operations is not trivial after all, and understanding how *each oper- ation* works requires making sense of the interactions between different types of elements within it.

Second, what exactly does it mean for a relationship to be 'harmoni- ous'? Harmony is a musical metaphor.[25] I have also talked about how actions 'fit together', but that is a mechanical metaphor, and actions are not parts of a machine, any more than they are musical notes.[26] I left this question unre- solved in my earlier publication. I confessed:

> It is difficult to be more precise in characterizing this quality of harmony in inter- operational interactions, or to reduce it to another, better-understood notion. We can go on listing synonyms: coordination, orchestration, concordance, back to coherence . . . It may be best to take 'harmony' (or 'harmonious') as a primitive in its meaning, and verifiable in the end only through the achievement of the aim of the activity. (Chang 2017b, p. 110)

---

[25] However, Liba Taub tells me that the musical notion of 'harmony' was probably itself a metaphor in the original Greek usage, drawing on the idea of things fitting together, as with the planks of a ship. There is another layer of meaning that Neurath's boat can take on, then!

[26] In Chapter 2 I will be criticizing the problematic uses of metaphors in 'correspondence realism', so I should be careful in wading into metaphors of my own!

But I was not quite satisfied with taking 'harmonious' as a primitive.

Third, what exactly is the relation between the success of an activity and its operational coherence? In my old paper I refrained from equating coherence and success, especially in order to allow for the possibility that a coherent activity could still fail due to some accidental extrinsic circumstance (and conversely, that an incoherent activity could succeed by accident). For example, I may do all my match-lighting operations sensibly, but be foiled by an unexpected gust of wind, a mischievous friend pouring a bucket of water all over me, or any number of other possible mishaps. If we can demarcate well enough the match-lighting activity itself from extrinsic accidents, then it would make sense to say that my match-lighting activity *is* coherent, but may occasionally be unsuccessful due to circumstances. But why exactly is it that operationally coherent activities should tend to be successful? What kind of mechanism or causal path might be involved in the production of success from coherence? I left this issue unresolved, too.

Coherence as Understanding and Coordination

Now, as it turns out, all of those puzzles about the meaning of operational coherence have one common solution, the germ of which is contained in the following statement from my old paper: 'A coherent activity *makes sense* in the realm of abstraction, but whether its actual execution is successful depends on all sorts of conditions. This is responsible for the sense that coherence and success are not synonymous' (Chang 2017b, p. 111; emphasis added). That is to say, operational coherence is a hermeneutical notion, concerning a pragmatic kind of understanding. What is *operationally* coherent is what makes sense for us to *do*, and 'sense' here is framed by our aims. But what does *sense-making* have to do with success? Surely I can't be suggesting that if an activity makes sense to me it will tend to be successful? The success of an activity is not *caused* by its coherence; rather, the coherence of an activity *consists* in doing what is sensible to do if one wants to succeed. Coherence is *design* for success, and that design is based on empirical learning: it makes *sense* to do what we think will succeed, and it doesn't make sense to do what we think is unlikely to succeed. Coherent activities are carefully designed so that they *would* work. The coherence–success relation is not one of cause and effect, but a hermeneutic–pragmatic act of sense-making in the context of purposive behaviour.

But what exactly does 'making sense' mean? There is nowadays a sizeable literature in the philosophy of science on the nature of understanding, to which I want to make links in the end.[27] However, initially I must set out in a different direction. This is because most of the extant literature is about the understanding of natural phenomena, or about our understanding of scientific theories. For my purposes here, it is necessary to consider first of all the understanding of our own actions, because that is what lies at the heart of operational coherence; this is what Oscar Westerblad (forthcoming) calls 'operational understanding'. In that case sense-making has to be approached partly from a psychological angle, as in Paul Thagard's much-neglected work on coherence. Thagard focuses on what the mind can hold together happily without too much dissonance, taking coherence as a relation between various types of elements: concepts, propositions, parts of images, goals, actions, etc. These elements can fit together (or not) through a variety of 'coherence relations', which may 'include explanation, deduction, facilitation, association, and so on' (Thagard 2000, p. 17). Using these relations we can make sense of how the different elements of our activities work together.[28] I should also look to the field of hermeneutics for a deeper understanding of understanding, but the traditional focus there is on the understanding of linguistic texts, so the connection to the understanding of actions will be indirect.

My own direct approach starts by taking operational coherence as something to do with rational action. If we start with the standard notion of instrumental rationality (means–end relationship), what makes sense for us to do is whatever will facilitate the achievement of our aims. But the usual treatments of instrumental rationality tend to exclude the hermeneutic dimension, seeing the means–end relation as basically causal. In my own initial thinking I was also projecting operational coherence onto the material dimension of actions, and that is why I could only talk in terms of metaphors. Things 'fitting together' is meaningless, unless there is a purpose under which the fit is judged; common images like the planks of a ship fitting together are deceptive, because in those cases the purpose (making the ship watertight) is taken for granted and not mentioned. The coherence of an activity is not some mysterious harmony between things in themselves, but it is a matter of how

---

[27] See esp. De Regt, Leonelli and Eigner (2009); De Regt (2017); Grimm, Baumberger and Ammon (2017); Grimm (2018); Stuart (2018).

[28] Thagard views coherence primarily as a matter of constraint-satisfaction, but I think many of his insights can be reworked in the direction of understanding. Interestingly, what is most akin to my thinking on operational coherence is Thagard's view of 'deliberative coherence' (operative in the realms of ethics and politics): 'if an action facilitates a goal, then there is a positive constraint between them' (Thagard 2000, p. 127). As for the problems of 'truth' and 'epistemic justification', Thagard considers them only to involve propositions as elements of coherence (ibid., p. 25, table 1). So his epistemology is not designed to deal with active knowledge, but I believe that it can be adapted to do so.

we bring together things and actions in order to achieve our aims. That is the sense in which operational coherence is **aim-oriented coordination**.

Some brief examples will be helpful in illustrating the pragmatic sense-making involved in operational coherence. There is a cartoon from *The Far Side* by Gary Larson showing a young man sitting up in bed in the morning looking attentively at a sign on his wall: 'Pants BEFORE shoes.' The joke is about someone needing to write that down to remind himself, but in all seriousness we do all have such rules lodged in our heads and in our bodies. Putting your shoes on before your pants[29] doesn't make pragmatic sense; you know that if you've tried it. Trying to go in opposite directions simultaneously, or trying to be in two places at the same time, certainly doesn't make sense for us, though you may think a bit differently if you are dealing with electrons in the quantum mechanical double-slit experiment. Trapping ultra-cold atoms with a laser is a coherent activity, because the physicists have learned to understand the conditions and operations that enable this feat. I may practise archery coherently, based on a certain sense of my own strength, of the properties of the bow and the arrow and the surrounding air, of the location of the target, and of the basic laws of mechanics.

To understand coherence, again it also helps to think about cases of incoherence. Some incoherence can be purely mental. In colloquial usage we often say 'incoherent' when someone is talking gibberish – not understandable, 'not even wrong'. But it is more interesting and informative to consider activities whose incoherence consists in a mismatch between how we think, what we want, what we do, and the way things are. Return for a moment to the example of archery: if the arrow does not hit the target, then that is a failure. Now I may say: oops, my hand slipped as I was stretching the bow; or, damn, there was this sudden gust of wind that I hadn't expected; and so on, to give myself and others an *understanding* of the failure. So, I can maintain the coherence of my activity in the face of isolated failures because my understanding of the whole activity remains intact. But if I keep missing the target completely and my failures are inexplicable, then it is incoherent to keep doing what I am doing. Then I must investigate, in the way that Peirce and Dewey say a disturbed situation gives rise to inquiry. I need to start thinking and doing things differently. Maybe I have to pull the bowstring harder, or revise the way I assess the amount of force exerted, or get my eyes checked to see if I am seeing the target well enough, or even adopt different laws of mechanics. If

---

[29] Larson means 'pants' in the American sense, meaning trousers. But the guideline would make even more sense with the British meaning of 'pants' (underpants)!

I keep doing what doesn't work instead of making such adjustments, then my activity is incoherent.

With the notion of operational coherence now framed in terms of pragmatic understanding, it becomes clear how the three puzzles that plagued my old definition of operational coherence can be solved. First, pragmatic sense-making can accommodate the coordination of heterogeneous types of elements, built around the central relationship between means and ends. Second, the elusive and metaphorical sense of 'harmony' is now reduced to the sense of pragmatic understanding. And finally, there is a positive relationship between the coherence of an activity and its success because we *write into* the design of a coherent activity our sense of what would succeed. We are safe from being 'untethered from reality' as long as our sense-making maintains a commitment to empiricism. Pragmatic understanding is not just 'in the head'. But what if people refuse empiricism itself, and insist that ignoring the lessons of experience *makes sense to them*? No amount of objectivist epistemology or self-righteous condemnation will stop such people; we can only win them over patiently by showing them the fruits of empiricism in the long run.[30]

Before I close the discussion of operational coherence, I should briefly address the difficulties involved in attributing coherence to other people's activities works. This is part of a general issue, which I have already touched on in Section 1.3 in relation to the identification of activities. I would argue that we are justified in interpreting the activities carried out by other agents as operationally coherent, even if they do not themselves articulate the aims and the coordination involved in their activities. Historians, including historians of science, confront this issue on a regular basis, as the past people who did not leave evidence of reasons for their actions cannot be further interrogated. It is a meaningful and instructive exercise to attempt an understanding of the activities of non-articulating agents as operationally coherent. I may be an 'idiot savant' of archery, who just somehow knows how to hit the target without much thinking or conscious planning; however, the astonished observers would attribute coherence to my activity if *they* could understand how my operations make sense in relation to the aim of hitting the target. This is not fundamentally different from how we deal with the general problem of other minds in practice – I make the decision to regard you as a conscious

---

[30] In not appealing to an absolute standard or authority for understanding, my view may be considered a relativist one. But relativism in the sense of rejecting absolutes is not a crude and bankrupt doctrine, as the imposing collection of recent works edited by Martin Kusch (2020) makes abundantly clear.

knowing agent with experiences that are similar to mine, even though I have no direct access to your experiences. This is just a reasonable way of living, based on a productive kind of respect.

The same may be done with animals. It is more natural to say that bees know how to collect nectar from the flowers and tell each other where the good flowers are, than to insist that they couldn't possibly possess knowledge because (we presume) they do not formulate beliefs in their tiny brains. I do think that spiders know how to build webs, squirrels know how to hide food (and even how to find it again, sometimes), dolphins know how to work together to corral fish, and migratory birds know how to navigate through thousands of kilometers of terrain with astonishing reliability. And AlphaGo surely knows how to play the game of go better than any human player. The attribution of knowledge in all these cases is based on our understanding of the agents' behaviours as operationally coherent activities. In daily life we rate such attributed knowledge highly and stand in awe of it. Why should epistemology ignore it? For my present purposes, however, it is enough that we make attributions or coherence, aims and knowledge in dealing with most humans and some very clear cases of other thinking beings. It is not necessary for me to engage in debates such as whether thermostats can be said to have consciousness and experience (Chalmers 1996, and responses to it).

1.5   Inquiry as Aim-Oriented Adjustment

In this section I take a dynamic view on active knowledge. That means looking into the nature of inquiry, the process through which we actively attempt to acquire and improve knowledge. I will elaborate on the idea that inquiry is the business of increasing operational coherence, a process of aim-oriented adjustment. In fully unrestricted inquiry, any aspect or element of the unsatisfactory initial situation facing the inquirers may be altered in order to bring about better operational coherence in their activities. Not only specific beliefs but capabilities and methods, and even aims themselves may be changed and improved in the process of aim-oriented adjustment. Various types of more restricted inquiry arise when certain elements are fixed. Starting from an examination of unrestricted inquiry helps us see the full range of processes of aim-oriented adjustment. Here I pay particular attention to inquiry processes that are usually neglected, including the crafting of concepts to handle novel experimental phenomena, the creation of new theoretical frameworks, and the adjustment of aims.

Aim-Oriented Adjustment

So far I have elaborated on the idea of active knowledge, whose quality consists in the operational coherence of epistemic activities. Now I want to take a dynamic view on the development of knowledge, and ask about the nature of inquiry. I start with a simple perspective: if good knowledge resides in operationally coherent activities, then the improvement of knowledge consists in the enhancement of operational coherence. Here I want to go clearly beyond the narrow image of inquiry as fact-finding, reorienting it and broadening it out into a picture of coherence-making. In Section 1.4 I stressed that operational coherence should be understood as a matter of pragmatic understanding; accordingly, I want to recognize the dimension of understanding in the process of inquiry, too. This view of inquiry is consonant with Peirce's and Dewey's mentioned in Section 1.1.

In Section 1.1, I stated that the enhancement of operational coherence was achieved through the process of **aim-oriented adjustment**, and I will now explain that notion in more detail. Let's start with a mundane example. Consider learning how to hammer a nail – how to make a tight grip on the handle, how to hit the nail on its head and how to recognize when you've done that, how to adjust the strength and number of the hammer-blows depending on the kind of wall, the kind of nail and other circumstances. Learning how to do this activity takes place through a process of trying out whatever it takes to improve the operational coherence of our activity in relation to the achievement of its aim. Generally, there is no pre-existing recipe for solving a real-life problem (rather than the kind of 'problem' that is an exercise laid out by the teacher who knows the answer already). Metaphorically, sometimes even literally, we have to twist and turn – try this and that, again and again, until something 'clicks' (or, until we settle into a comfortable and effective routine). This is quite similar to what Ludwig Fleck calls 'tuning'. As Andrew Pickering (1995, p. 121) puts it: 'The scientists tried varying the prototype recipe [for the Wassermann test for syphilis] in all sorts of ways and eventually arrived at a recipe that was medically useful.' We reach a bodily-and-conceptual understanding of the task as we go. Even learning to *see* is like this, and likewise for other modes of perception, too, as Alva Noë (2004) has vividly emphasized. Learning various second-person interactions, whose importance I emphasized in Section 1.2, also works largely by such aim-oriented adjustments: handshakes, explanation, co-authoring, telling jokes, boxing, ballroom dancing and moving furniture are just a handful of randomly selected examples.

It is instructive to revisit the classic example of learning to ride a bicycle, so memorably discussed by Polanyi. Initially the novice does not know how to keep himself from falling, and doesn't understand how that is done. The helpful older sister gives him tips like 'turn into the direction in which you are beginning to fall', but this advice makes no sense to the boy in the abstract, and when he tries to put it into practice, it doesn't work. However, trial and error eventually shifts something in the brain and the muscles, and he is able to ride, wobbly as he may be; at this stage the thing about turning into the direction of falling starts to make that conceptual–bodily sense. As his skill improves, he also begins to understand things like how a slight turn can be achieved by the shifting of body-weight without turning the handle. Through such learning-how-to improvements, his bicycle-riding continues to increase its operational coherence. If I ask myself whether what I am doing makes sense while riding my bicycle, I would say 'yes' in both ways: the way I move my muscles and shift my weight around makes inarticulate sense to me as I do it, and I can also understand what I'm doing by articulating it and putting it into practical rules like 'shift the body weight to make a slight turn'.

Unrestricted and Restricted Inquiry

In order to get a general sense of how inquiry works, I believe we must begin with a view of unrestricted inquiry, only afterwards asking how its character is modified when particular restrictions are placed on it. It is important to keep in mind that any elements of the problematic situation that gives rise to inquiry may in principle be revised or discarded, and new elements may also be introduced. The open-endedness of unrestricted inquiry is typical of what Kuhn ([1962] 1970) vividly described as 'extraordinary research', which often brings about revolutionary change. In such inquiry everything is subject to change, including presuppositions, methodology, aims, criteria of judgement, and the list of significant problems. In contrast, much of what is normally recognized as inquiry (or research) is of a much more restricted type (as in what Kuhn considers 'normal science'), and can only take place on the basis of some unrestricted inquiry that has previously been carried out. As an example of a most restricted type of inquiry, consider what we might call fact-gathering. Here the aims and questions and methods are all fixed, except for the actual contents of the blanks to be filled by well-regulated acts of observation. Fact-gathering is the simplest kind of inquiry that generates propositional knowledge as an outcome, but it would be a mistake to regard it as the prototype or paradigm of all inquiry. On the contrary, the successful execution of *unrestricted* inquiry is the foundation of all cognitive activity, because outcomes of unrestricted inquiry lie at the foundation of language-use, mathematics, experimental design, causal reasoning, theoretical explanations, and almost all other aspects of intelligent life. In more restricted types of inquiry so much is already settled, and the truly challenging and exciting unrestricted stages of the inquiry process tend to be hidden from view.

Even though Kuhn was wrong to imply a sharp dichotomy between 'normal' and 'extraordinary' research, his distinction still makes perfect sense if we think of it as pointing to the two ends of a spectrum. Where I really want to depart from Kuhn's perspective is in regarding 'normal' science as the *normal* state of science, or even definitive of science.[31] On the contrary, various kinds of restricted inquiry are only results of restrictions that are placed on inquiry as temporary and local expedients. Although in popular imagination Kuhnian extraordinary science is only associated with field-changing scientific revolutions, Kuhn himself did acknowledge that extraordinary research can happen at any scale. For example, the discovery of X-ray was a small-scale change but the process had the same character as the large-scale revolutions (Kuhn [1962] 1970, pp. 92–3). Kuhn's conception of normal science as 'research under a paradigm' expresses very well the restricted and well-prescribed character of much of mainstream inquiry in a scientific field. In contrast, extraordinary research is what scientists engage in when they feel that it is necessary to depart from the ruling paradigm in order to solve an urgent puzzle.

Recognizing unrestricted inquiry as the basic form of inquiry allows us to recognize that inquiry results in the development of all aspects of active knowledge, many of which have been neglected by philosophers. Within philosophy of science, attention has been limited mostly to several specific well-controlled types of inquiry: fact-gathering, hypothesis-testing, classification, theory-construction, the construction of theoretical explanation, and the development of observational methods. In the rest of this section I want to give some attention to some important aspects of inquiry that are not often discussed by philosophers of science, and bring out and highlight the pertinent processes of aim-oriented adjustment.

---

[31] See Kuhn's statement in his legendary debate with Karl Popper: 'It is normal science, in which Sir Karl's sort of testing does not occur ... which most nearly distinguishes science from other enterprises. If a demarcation criterion exists ... it may lie just in that part of science which Sir Karl ignores ... In a sense, to turn Sir Karl's view on its head, it is precisely the abandonment of critical discourse that marks the transition to science' (Kuhn in Lakatos and Musgrave 1970, p. 6).

The Making of Experimental Meaning

In experimentation, the most unrestricted inquiry takes place when new kinds of phenomena are discovered and an intense discomfort develops in not knowing how to make sense of them. Then a nearly desperate sort of aim-oriented adjustment takes place, to create new meaning in the new phenomena. A good example is how scientists two centuries ago tried to make sense of the electromagnetic effect discovered in 1820 by Hans Christian Ørsted (see Figure 1.1). A metallic wire is laid above a magnetic needle in a direction parallel to the needle; the magnetic needle turns when the wire is connected to a battery and a current of electricity flows through it. Factually, there was no difficulty concerning Ørsted's result – the phenomenon was clearly observed, easily replicated, and never seriously disputed. The challenge was how to make sense of it: why did the flow of electricity have a *magnetic* effect, and why did the electric current push the magnetic needle in a direction *perpendicular* to its

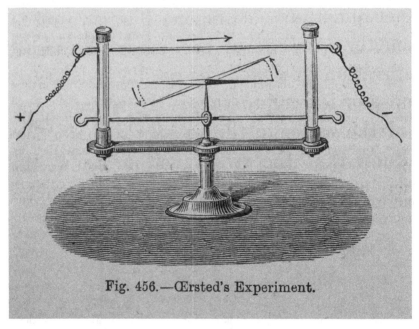

Fig. 456.—Œrsted's Experiment.

Figure 1.1   An experimental arrangement demonstrating the electromagnetic effect first discovered by Hans Christian Ørsted, from Privat-Deschanel (1876), p. 656, fig. 456.
Courtesy of the Whipple Library, Cambridge.

own flow (rather than sweeping it along, as it were)? No mechanism for such an effect could be found within the dominant Newtonian scheme of forces between point-particles acting along a straight line connecting them.

In trying to understand how scientists tried to cope with this situation, I draw my lessons from Friedrich Steinle's ideas about 'exploratory experimentation', which were in large part developed through his study of how André-Marie Ampère and Michael Faraday developed their knowledge of electromagnetism (Steinle [2005] 2016). In exploratory experimentation, systematic high-level theory is absent and researchers start by focusing on discovering empirical regularities. The initial method employed is the systematic variation of all known experimental parameters that are presumed to be relevant. Sometimes this works out well, but when it does not, 'researchers consider the possibility that the failure might have something to do with deficiencies in the basic concepts of the field and thus feel encouraged to form and introduce new concepts that capture experimental findings' (ibid., p. 314). There is no algorithm to follow in that creative process, so the twisting-and-turning of aim-oriented adjustment enters the scene. Steinle points out that Ampère invented the concept of 'right and left of current' and that of a current circuit through his attempt to find the right concepts with which to express empirical laws about the electromagnetic effect.

Steinle builds on David Gooding's sadly underappreciated work on embodied agency and practical thinking in scientific work, which culminated in his book *Experiment and the Making of Meaning* (1990). Gooding follows how Faraday conceived concentric circles of magnetic influence around the current-carrying wire, in planes perpendicular to the direction of the current. Rather than adopting the approach dominant in France (led by Ampère and others) to reduce electromagnetic phenomena to Newtonian-style forces as much as possible, Faraday drew inspiration from his mentor Humphry Davy, who had placed magnetic needles on the plane perpendicular to the wire and saw them turn as to form a ring (Gooding 1990, p. 53). The same sort of ring pattern could also be shown with iron filings spread on a cardboard piece. Faraday began to see his 'lines of force' filling the space around electrically and magnetically active bodies; this idea later developed into the concept of fields in the hands of James Clerk Maxwell and others.

But through what process did Faraday come up with such concepts? Faraday's inquiry was a process of aim-oriented adjustment, employing various cognitive means to achieve the aim of explaining and extending electromagnetic phenomena. Based on his minute examination of Faraday's laboratory notebooks and other pertinent archival sources, Gooding concludes: 'Faraday is thinking *through* doing as well as *about* doing. Some of these thoughts are inherently ambiguous until articulated into configurations of real or imagined entities (images, models or concrete apparatus).' Faraday's creative thinking was a full blend of visual imagery, tactile sensation, reflections on laboratory observations, and the invention and use of new experimental apparatus. His early electromagnetic experiments were mostly *not* tests of well-conceived hypotheses: 'Faraday was experimenting to *realize* possibilities, not to *decide between* two distinct or incompatible interpretations' (Gooding 1990, p. 124; emphases original). Through continually evolving experimentation Faraday articulated previously unknown phenomena and created new meanings. At the same time he also created wondrous new apparatus like his famous rotation devices (see Figure 1.2), in which a vertically suspended current-carrying wire is made to rotate around the pole of a magnet underneath it (or a magnet is made to rotate around a wire), in a pool of mercury. Such devices served as embodiments of Faraday's new ideas. Through this line of inquiry Faraday not only opened the path to the establishment of modern technological civilization, but created new understanding embodied in a set of new operationally coherent activities.

Establishing New Theoretical Frameworks

The creative enhancement of operational coherence through aim-oriented adjustment happens in more theoretically focused inquiry, too. For example, consider the physicists of Einstein's generation grappling with the puzzle of the Michelson–Morley experiment (for full details see Holton 1969; Miller 1981; and Staley 2008). Assuming that light is a wave in the aether and the earth is moving around in the aether, the apparent speed of light should depend on the earth's motion, but there was no detectable variation that could be found. It was certainly not obvious what needed to be fixed in this situation in order to bring it into a state of harmony. Many ingenious accounts were given in order to explain why the motion of the earth through the aether would be undetectable, ranging from the idea that the earth dragged around the aether in its vicinity to the systematic changes in the observations of time and space coordinates proposed by Hendrik Antoon Lorentz (and similarly by George FitzGerald). Einstein's solution was more daring: he got rid of the problematic situation altogether by proposing his two postulates: the principle of relativity, and the principle of the constancy of the velocity of light (regardless of the motion of the source or the observer of light). With these postulates Einstein set about reconceiving the very concepts of space, time, mass and energy. Einstein showed that a very coherent system of practice could be built on this initially implausible basis, a system that included the activities of defining

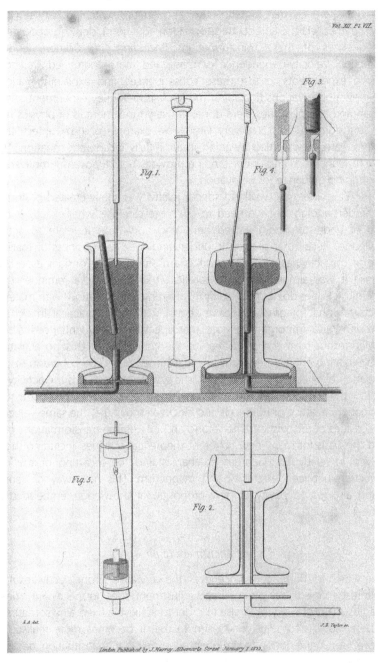

Figure 1.2   Faraday's compact rotation device, from Faraday (1822), plate 7.
Courtesy of Cambridge University Library.

frames of reference, transforming physical quantities switching between reference frames (via the Lorentz transformation equations), deriving observable consequences of these transformations (including time-dilation, length-contraction, velocity-dependence of mass, and mass–energy equivalence), devising experiments to test these consequences, and explaining various well-known observations (including the Michelson–Morley experiment). After Einstein and his colleagues were done, the very fundamentals of physics had been transformed. This is a very large-scale example of how unrestricted inquiry creates new settled meaning in an initially incoherent situation. The desire for coherence and understanding provides both a powerful motivation and ongoing guidance for innovation.

Einstein's 1905 work on special relativity certainly created profound new understanding, but we need to ask more carefully: what exactly is the sense of understanding involved here, and how does it relate to active knowledge? Recall that pragmatic understanding is the business of making sense of our activities. Einstein's work is a useful case to consider especially because it was almost entirely theoretical; this serves as a reminder that epistemic activities do not necessarily involve physical operations with material objects, and that 'pragmatic' does not equal 'material' or 'practical' in a crude sense. It is also important to note that special relativity enhanced active knowledge not so much by increasing the coherence of existing activities, but by creating entirely new activities that turned out to be coherent. At the foundation of all activities in the relativistic system of practice is the activity of defining the frame of reference for any given observer, by putting in (by imagination) a lattice of rigid rods and clocks all moving at the same velocity. The activities of relativistic frame-setting and inter-frame transformations provided the basis for many new coherent theoretical activities, including those involved in relativistic quantum mechanics and the recasting of the old connection between electricity and magnetism. This new way of sense-making also led to plenty of new propositional knowledge, embedded in the new activities.

The Adjustment of Aims

In the search for better coherence, even the very aims of one's activities may be altered. That is to say, aim-oriented adjustment does not necessarily mean the adjustment of everything else in order to achieve a fixed aim; sometimes the best move is to adjust one's aim so that it becomes more realistic to achieve. I retain the notion of aim-orientedness even in that situation, because the operational coherence of an activity is still defined in relation to the aim of the activity at each moment, even though the situation is dynamic because our aims are not fixed in the long term. Aims can and should change in the process of inquiry, if we learn that what we were trying to do is not plausible and the best way to enhance coherence is to try for a different aim. For example, while engaged in the activity of hammering nails, we may find ourselves surrounded by steel walls; the reasonable thing to do in that situation would be to stop hammering, and find some other way of hanging the picture. It makes sense to stop trying to achieve an unfeasible aim. If we change the inherent aim of an activity, the activity ceases to exist. With a new aim we enter into a new activity, with a whole new judgement of coherence. And more positively, what we do well can often become something that we *want* to be doing, around which we also build other activities.

Talk of hammer and nail may sound idle in the context of the philosophy of science, but the modification of aims in search of coherence also happens in serious scientific practice. The case of Einstein and special relativity again provides an excellent illustration. Einstein did not solve the original problem of accounting for the result of the Michelson–Morley experiment (and other 'aether-drift experiments') in terms of the motion of the earth through the aether. Rather, he dismissed that problem by declaring the aether 'superfluous', and launched whole a new set of epistemic activities. His proposal was attractive to many other physicists who had become weary of their inability to solve the original problem. The course of inquiry can very well lead us to a reassessment of what it is that we ought to be aiming to know, and what questions are worth investigating.

1.6 Pragmatism and Active Knowledge

My thinking in this book has been strongly inspired by the pragmatist tradition of philosophy, so it is important that I explain what I take pragmatism to be, and why it is so relevant to thinking about active knowledge. Pragmatist philosophy is often concerned with clarifying the meanings of concepts in terms of their practical implications. More broadly, however, pragmatism is a philosophy that concerns itself with the nature of practices, which is therefore suitable as a framework for understanding the nature of active knowledge. Pragmatism as I see it is a thorough and relentless empiricism, which insists that experience is the only ultimate source of any kind of learning, and takes experience in its full sense as something that active knowers undergo. For pragmatists, empirical learning includes the learning of methods, all the way to the choice of appropriate logical axioms. Such a pragmatist perspective will helpfully inform all the other discussions in the remainder of this book. My interpretation of pragmatism is also offered as a small contribution to the recent revival of pragmatist philosophy in various quarters.[32]

Beyond Semantics: Pragmatism as a Philosophy of Practice

What is pragmatism, and what are its implications for the philosophy of science? A very good definition actually comes from an ordinary dictionary (*Webster's Ninth New Collegiate Dictionary*, 1986): 'an American movement in philosophy founded by C. S. Peirce and William James and marked by the doctrines that the meaning of conceptions is to be sought in their practical bearings, that the function of thought is to guide action, and that truth is preeminently to be tested by the practical consequences of belief'. The first part of this definition is nothing but a version of Peirce's 'pragmatist maxim', paraphrased by James here ([1907] 1975, p. 29): 'To attain perfect clearness in our thoughts of an object, then, we need only consider what conceivable effects of a practical kind the object may involve – what sensations we are to expect from it, and what reactions we must prepare.'[33] The Peirce–James pragmatist maxim naturally led to a semantic interpretation of pragmatism, which is perhaps the dominant interpretation today. According to Catherine Legg and Christopher Hookway (2021, sec. 2), the pragmatist maxim 'offers a distinctive method for becoming clear about the meaning of concepts and the hypotheses which contain them. We clarify a hypothesis by identifying the practical consequences we should expect if it is true.' In this way, pragmatism shares much with operationalism, and with the verificationism that was widely taken as a core doctrine of logical positivism.

James presented pragmatism as a 'method for settling metaphysical disputes that otherwise might be interminable' (James [1907] 1975, p. 28). A dispute is idle unless some 'practical difference' would follow from either side being correct. James opened his lecture on 'What pragmatism means' with an apparently trivial anecdote: on a visit to the mountains, his friends got into a 'ferocious metaphysical dispute' – about a squirrel! The animal was hanging on to one side of a tree trunk, with a human observer on the other side:

---

[32] The literature is too large and varied for me to survey adequately here. I will be discussing various individual authors' works in the remainder of the book. Some excellent collections of recent works include volumes edited by Cheryl Misak (2007a), Roberto Frega (2011) and Kenneth Westphal (2014).

[33] For the original formulation, see Peirce (1878).

This human witness tries to get sight of the squirrel by moving rapidly round the tree, but no matter how fast he goes, the squirrel moves as fast in the opposite direction, and always keeps the tree between himself and the man, so that never a glimpse of him is caught. The resultant metaphysical problem now is this: *Does the man go round the squirrel or not?*

And here is James's solution of the problem:

[The correct answer] depends on what you *practically mean* by 'going round' the squirrel. If you mean passing from the north of him to the east, then to the south, then to the west, and then to the north of him again, obviously the man does go round him, for he occupies these successive positions. But if on the contrary you mean being first in front of him, then on the right of him, then behind him, then on his left, and finally in front again, it is quite as obvious that the man fails to go round him, for by the compensating movements the squirrel makes, he keeps his belly turned towards the man all the time, and his back turned away. Make the distinction, and there is no occasion for any farther dispute. (Ibid., pp. 27–8; emphases original)

In this manner, the 'pragmatic method' promises to eliminate all irresolvable metaphysical disputes, and rather more important ones than the squirrel-puzzle, too.

Even though I completely endorse the semantic tradition of pragmatism that James's squirrel example would seem to embody, my own emphasis is different. I follow Philip Kitcher's (2012, pp. xii–xiv) warning against the 'domestication' of pragmatism. Focusing on semantics can be a very effective method of domestication, making pragmatism look like a rather innocuous and interesting variation on normal analytic philosophy. I want pragmatism to be a philosophy that helps us think better about how to do things, not just about what our words mean in relations to actions. Recall the second part of the dictionary definition of pragmatism quoted above: 'the function of thought is to guide action'. Concerning the squirrel, one might wonder if what James advocates isn't just a matter of defining one's terms carefully. But I think that the sort of disambiguation offered by James is tied closely to potential practical ends. If my objective is to make a fence to enclose the squirrel, then I have gone around the squirrel *in the relevant sense*; if my objective is to check whether the wound on his back has healed, then I have failed to go around the squirrel in the relevant sense. It is the pragmatic purpose that tells us which sense of 'going round' we *ought* to mean. Semantics should be a tool for effective action. This is fully compatible with Huw Price's (1988) neo-pragmatist functionalism about truth and other notions.

Pragmatism as Relentless Empiricism

One very important reason why people often do not like to go beyond the semantic dimension of pragmatism is the fear of what happens if we go further and adopt the pragmatist theory of *truth*.[34] This issue needs to be tackled head-on. It is crucial that we reject the common misconception that pragmatism takes whatever is *convenient* as true. The 'pragmatic theory of truth' attributed especially to James is widely regarded as absurd, and this has contributed greatly to the disdain for pragmatism among tough-minded philosophers. Here is probably the most notorious statement by James:

> 'The true,' *to put it very briefly, is only the expedient in the way of our thinking, just as* 'the right' *is only the expedient in the way of our behaving.* Expedient in almost any fashion . . .

I think James's choice of the word 'expedient' here was unfortunate, as sounding too much like just 'convenient' or 'useful'. Or perhaps the word had quite a different connotation back then; that is for James scholars to debate. At any rate, the statement actually continues as follows:

> . . . and expedient in the long run and on the whole of course; for what meets expediently all the experience in sight won't necessarily meet all farther experiences equally satisfactorily. Experience, as we know, has ways of *boiling over*, and making us correct our present formulas. (James [1907] 1975, p. 106; emphases original)

In my view, what this passage really shows is James the staunch empiricist, declaring that the source of truth is experience, and that it is *futile to entertain any more grandiose notion of truth*. (I will have more to say about the pragmatist theory of truth in Section 4.6.) This provides an important clue to my interpretation of pragmatism, which is to understand it as a thoroughgoing, complete and relentless empiricism. Empiricism recognizes experience as the only ultimate source of learning, and refuses to acknowledge any higher epistemic authority. Here is James again ([1907] 1975, p. 31):

> Pragmatism represents a perfectly familiar attitude in philosophy, the empiricist attitude . . . both in a more radical and in a less objectionable form than it has ever yet assumed. A pragmatist turns his back resolutely and once for all upon a lot of inveterate habits dear to professional philosophers. He turns away from abstraction and insufficiency, from verbal solutions, from bad *a priori* reasons,

---

[34] My statement may be puzzling to those who treat semantics as a matter of truth-conditions. I prefer to take semantics as a study of meaning in a broader sense.

from fixed principles, closed systems, and pretended absolutes and origins. He turns towards concreteness and adequacy, towards facts, towards action, and towards power. That means the empiricist temper regnant, and the rationalist temper sincerely given up.

According to pragmatism as I see it, how philosophy serves life is by paying full and thorough attention to the experiences that constitute life.

In empiricism as it is generally presented in philosophy these days, the view of experience presented is extremely limited, seen as a matter of gaining information through sense-perception. Hans Radder (2006, ch. 2) rightly laments the 'absence of experience in empiricism'. The classical pragmatists had a feeling for the richness of experience, which I think all empiricists should recover. Cheryl Misak (2013, p. 12) argues that the early pragmatists were inspired by Ralph Waldo Emerson, who wanted empiricism, but not 'paltry empiricism'; for Emerson, experience of course included emotional and passional experience. James's 'radical empiricism' involved paying respectful attention even to religious, mystical and parapsychological experiences. It is an important part of pragmatism to take 'experience' as the full lived experience of human beings, recognizing its full range and all of its aspects. Pragmatism also understands experience in the context of action, which goes well with my conception of active knowledge. In taking such a well-rounded view of experience, pragmatism can look surprisingly different from what philosophers normally mean by 'empiricism'.

Let us consider further what a full understanding of experience involves. Empiricism is significantly perverted when it is taken to imply that we should assign absolute epistemic authority to results of sense-perception, ignoring other dimensions of experience. And sense-perception itself is much more complex than often imagined by many philosophers of science, involving much more than 'the five senses'. We have a great deal to learn from the phenomenologists in this regard, and there is much potential in a synergy between phenomenology and pragmatism.[35] Proprioception and muscular tension are inevitable underlying ingredients to all sensation, even when they are not subjects of conscious experience. Hacking (1983, p. 189) brings this consideration into scientific observation, too: 'you learn to see through a microscope by doing, not just looking'. He invokes George Berkeley's *An Essay Towards a New Theory of Vision*, 'according to which we have

---

[35] We could do worse than starting again with Maurice Merleau-Ponty ([1945] 1962), and then moving on to current authors such as Alva Noë (2004) and Mazviita Chirimuuta (2015). Long ago Herbert Spiegelberg (1956) noted an obvious affinity to phenomenology in the works of James, and made an in-depth comparative study of Peirce and Husserl.

three-dimensional vision only after learning what it is like to move around in the world and intervene in it'. Berkeley also points out that muscular sensations in the eye are an essential part of the experience of seeing (Berkeley [1709] 1910, pp. 16–20). Taking experience as it is really lived also means paying more attention to the active dimensions of experience, which is also to say that we focus on the *practices* of perception and observation. Even just looking at something is a concerted activity undertaken with a purpose. A focus on action also means a focus on the active agent, with a full awareness of the nature of epistemic agents as discussed in Section 1.2. We should also recognize experience as process. James's discussion of the 'stream of consciousness' referred to Henri Bergson's ideas on the passage of time, 'real duration', and memory (Bergson [1896] 1912). We need to take experience as a process of life, not as a set of statements describing contents of sense-perception. Common views on the nature of observation, experiment and empirical evidence need significant updating and revitalization.

The pragmatist view of knowledge is deeply rooted in humanism. Here I am using the term 'humanism' broadly, in the spirit of the now-forgotten German-British philosopher Ferdinand Canning Scott Schiller (1939, pp. 65–80). Humanism concerning science recommends that we understand and promote science as something that people do, not as a body of knowledge that exists completely apart from ourselves and our investigations. Science is a thoroughly human activity even when it is aimed at the production of the most abstract and impractical kind of knowledge. Empiricism itself can be seen as a form of humanism: ultimately, the only source of learning we have is human experience, shaped by human nature. As James famously put it ([1907] 1975, p. 37): 'The trail of the human serpent is thus over everything.' Perhaps this sort of humanism is not such a controversial stance (with its roots going back at least to Kant), but I think there is much value in spelling out its meaning and implications carefully. The spirit of humanism has been summarized in another way, and rather poetically, by Clarence Irving Lewis in his review of Dewey's 1929 masterpiece, *The Quest for Certainty*:

> Man may not reach the goal of his quest for security by any flight to another world – neither to that other world of the religious mystic, nor to that realm of transcendent ideas and eternal values which is its philosophical counterpart. Salvation is through work; through experimental effort, intelligently directed to an actual human future. (Lewis 1930, 14)

This passage is especially nice because it brings together the two pragmatist philosophers that I have found most inspiring, and ties together the humanist and the empiricist strands of pragmatism. Active knowledge fits very well into this empiricist-humanist spirit of pragmatism.

The most important thing about humanism as I see it is not the focus on the biological species *Homo sapiens*. Humanism is not incompatible with attention to artificial intelligence, animal cognition, or extraterrestrial intelligence. On the contrary, our understanding of non-human cognition would only be enhanced by a comparative view based on a full understanding of human cognition, and vice versa. The most fundamental point about pragmatism, as I take it, is that knowledge is created and used by some sort of epistemic agents, namely intelligent beings who engage in actions in order to live better in the material and social world. Recall Dewey's trenchant critique of the 'spectator theory of knowledge' (see Kulp 2009).

An Empiricist View on Methodology and Logic

The staunch empiricism at the core of pragmatism also encompasses methodology. In Dewey's view, method-learning is an empirical process as much as any other learning: 'through comparison-contrast, we ascertain *how* and *why* certain means and agencies have provided warrantably assertible conclusions, while others have not and *cannot* do so in the sense in which "cannot" expresses an intrinsic incompatibility between means used and consequences attained' (Dewey 1938, p. 104). Methodological rules are contingently generalizable, just like any general empirical statements: 'inquiry, in spite of the diverse subjects to which it applies ... has a common structure or pattern ... applied both in common sense and science' (ibid., p. 101). There is no special method for method-learning, which is just a kind of empirical learning from experience, our assumptions being conditioned by success. And the success of methodology is merely a matter of being 'operative in a manner that tends in the long run, or in the continuity of inquiry, to yield results that are either confirmed in further inquiry or that are corrected by use of the same procedures' (ibid., p. 13).[36] Dewey held strongly to the continuity of rules – of everyday inquiry, scientific method, and even logic (ibid., pp. 4–6), all of which arise from successful habits of thinking (ibid., p. 12). It was not a category-mistake that Dewey titled his last great work (published when he was nearly eighty) *Logic*, and subtitled it *The Theory of Inquiry*.[37]

---

[36] Dewey also says success is measured by 'coherency' but doesn't seem to say what exactly *that* means.
[37] For a thorough treatment of Dewey's ideas on logic and methodology of science, see Matthew Brown (2012).

Dewey brings methodology down to earth: 'we know that some methods of inquiry are better than others in just the same way in which we know that some methods of surgery, farming, road-making, navigating or what-not are better than others' (1938, p. 104). Ways of reasoning have also developed through the course of the history of science, sometimes in quite fundamental ways, and often it is the development of very specific new methods that challenge the accepted general rules of thought. Looking back at the history of science, we can spot disputes about the best ways of reasoning in almost every period. Should we think of velocity and weight as quantities? The medieval Aristotelians had thought not, but they were defeated by the indisputable successes of quantification (Crombie 1961). Are thought-experiments legitimate ways of generating and supporting physical theories (Stuart, Fehige and Brown 2018)? Galileo, and later Einstein, thought so, and others disagreed. How about computer simulations (see, e.g., Galison 1997, ch. 8)? Such debates and changes are still ongoing: witness the dispute surrounding whether mathematical proofs done by computers should be considered valid (Burge 1998), or more practically, how best to put to use medical diagnoses made by artificial intelligence systems (Ahmad et al. 2021). Less outlandishly: which is the right sort of statistics to employ in empirical testing, frequentist or Bayesian (Mayo 1996)? What are the circumstances in which a randomized controlled trial is the appropriate method of hypothesis-testing (Cartwright 2011; Cartwright and Hardie 2012)?

It is an indisputable historical fact that science has learned new and better methods of inquiry in the course of its development. And the adoption of a new method is often accompanied by changes in fundamental epistemic standards as well. Take Alan Chalmers's discussion of Galileo's establishment of the telescope as a trusted means of making astronomical observations, supe-rior to naked-eye vision (Chalmers 2013, pp. 151–5). This was a momentous event that had a deep implication reaching far beyond itself, as it was one of the key early instances in which the verdict of an instrument was deemed to be epistemologically superior to human sensation. I have told a similar story concerning the overriding of the sensation of hot and cold by thermometer-readings (Chang 2004, p. 47). These methodological changes were not sanc-tioned by some super-method or an overarching theory of physics, but only by some detailed case-by-case arguments about how the telescopic or thermo-metric observations should be given more credence than unaided perception. For example, the telescope showed some details that naked-eye observations did not show (such as the phases of Venus and the craters on the moon), and these details also made sense (if one adopted the Copernican theory, that is). Galileo also pointed out some inconsistencies in the human visual perception of the brightness and shape of planets and stars. Galileo was not able to argue from any first principles that the telescope should be granted epistemic authority (especially in the absence of a well-developed theory of optics at the time), but the use of telescopes in some concrete inquiries worked out very coherently.

It would have been the ultimate prize for a pragmatist to argue successfully that even logic was only pragmatically justified, rather than being a set of eternally valid 'laws of thought'. If so, the methods of science and any other rules we have in life are *of course* going to be revisable and evolvable. I think that is exactly what Dewey aimed to achieve. Logical rules are decreed to be true when you are in logic class, but when logic is employed in other settings it must be judged by its fruits, in terms of how well it supports operationally coherent activities. Logic for Dewey is still a normative discipline, about how we *should* think in order to think well. But the normative force of logic ultimately rests on its empirical success, not on any a priori requirements. For example, consider the cogency of the 'the principle of explosion' concerning material implication, which dictates that from a contradiction one can deduce any proposition. Is that really a productive way of thinking? I certainly have trouble thinking of any real-life situation in which we should go wild and believe any propositions at all, just because we have a contradiction on hand.

Dewey denies that logic exists at all apart from the subject matter of reasoning: 'all logical forms (with their characteristic properties) arise within the operation of inquiry and are concerned with the control of inquiry so that it may yield warranted assertions' (Dewey 1938, pp. 3–4). What people normally consider 'logic' is only the most general among the rules of thought that have been shown to be good, most abstractly expressed. In Dewey's view logic is a historical and empirical discipline in which we continually learn better ways of reasoning. Declaring that logic is a 'progressive discipline' (ibid., p. 14), he speaks of 'the needed reform of logic' (ibid., title of chapter 5) which should be based on a full historical awareness. For example, Dewey argues that Aristotelian logic was a system admirably suited for the science and philosophy of ancient Greece, but no longer suitable (ibid., pp. 82–93). As key elements of Aristotelian thinking that have been abandoned, he lists: essentialism, the emphasis on quality over quantity, static classification as the form of knowledge, the heterogeneous and hierarchical structure of the universe (again, one can see how logic and methodology blend into each other in Dewey's thinking). He faults his contemporary logicians for tending to retain the form of classical logic while abandoning the metaphysical and operational underpinnings of it.

One might think that Dewey was just a 'fuzzy' thinker who was ignorant about logic as practised by professional logicians. But no one could level the same accusation against C. I. Lewis, whose views on logic were very similar to Dewey's. Lewis had built a great and lasting reputation in the area of symbolic logic (for which he is still remembered) before he became known as a pragmatist. Lewis (1929, p. vii) himself stated that his pragmatist epistemology had in fact originated from his work in symbolic logic. Early on Lewis felt, like Dewey, a dissatisfaction with certain features of the fundamental notion of material implication in classical logic. Founded on the principle of excluded middle, classical logic dictates that P (materially) implies Q whenever it isn't the case that P is true and Q is false. Lewis thought that this notion did not correctly express what we mean by 'implication' in natural language, and it should be replaced by what he called 'strict implication', which did not have undesirable consequences like the principle of explosion that Dewey deplored. But critics pointed out that Lewis's strict implication had its own awkward features, and he was forced to admit that each system of logic had various virtues and vices. This pushed Lewis towards pluralism about logic, and about conceptual frameworks in general.[38]

There *are* different systems of logic, and anyone who wants to reason logically must start by adopting a particular system of logic. But the only plausible and non-arbitrary way of justifying the choice of a logical system would be on pragmatic grounds, because appealing to the rules of logic in arguing for this choice would clearly be question-begging. So it may actually turn out that the treatment of logic is the most convincing part of pragmatism! Lewis summed up his view as follows: 'the choice of conceptual systems for the interpretation of experience ... is a matter of pragmatic choice, whether that choice be made deliberately, or unconsciously and without recognition of its real grounds' (Lewis 1929, p. 300). Looking at the current proliferation of non-classical (or alternative) logics and the successful application of some of them in the design of intelligent systems, I think we must admit that Dewey and Lewis have been vindicated. The spirit of Dewey's work lives on in current work on 'anti-exceptionalism' about logic, as expressed here by Ole Thomassen Hjortland (2017, p. 631): 'Logic isn't special. Its theories are continuous with science; its method [is] continuous with scientific method. Logic isn't a priori, nor are its truths analytic truths.

---

[38] On Lewis's development see Schilpp (1968); Rosenthal (2007); Misak (2013), ch. 10; Stump (2015), ch. 5.

Logical theories are revisable, and if they are revised, they are revised on the same grounds as scientific theories.'[39]

To sum up: a truly empiricist philosophy of science, as I take pragmatism to be, should recognize clearly that scientific inquiry is itself a kind of human experience. According to the pragmatists inquiry is pervasive in life, as well as scientific practice; it could even be that experience is inherently inquisitive. And not only does inquiry engage with experience, but the process of inquiry itself is a type of experience. Learning from experience includes *learning from the experience of learning*. Philosophers need to pay attention to the processes of knowledge-production and knowledge-use, and ask how we can best organize and support the epistemic activities involved in those processes. If we conceive pragmatism as a philosophical commitment to engage with practices, pragmatist epistemology should concern itself with all practices relating to knowledge. And I believe that this is something that the classical pragmatists were seriously engaged in.

In the remaining chapters of this book I will carefully unfold the implications of pragmatism on epistemology and metaphysics, especially in the context of the philosophy of science. In this chapter I have focused on the pragmatic nature of knowledge and presented a pragmatist view of the process of inquiry. Central to my thinking throughout will be the notion of *operational coherence*, which I have defined in Sections 1.1 and 1.4. Operational coherence is the anchor of the kind of realism that pragmatists (and empiricists in general) can embrace, as I will explain fully in Chapter 5. Any sort of realism needs to take something real as the object of knowledge, so in Chapter 3 I also advance a pragmatist notion of reality, which is also based on the notion of operational coherence. Realism also involves the basic idea that through our inquiries we can learn truths about realities, and I will propose an updated pragmatist notion of truth in Chapter 4.

---

[39] Hjortland traces this view to Quine, Maddy and Priest among others. For an introduction to non-classical logics see Priest (2008) and Gottwald (2020).

CHAPTER 2

Correspondence

2.1 Overview

Correspondence Realism: Just Common Sense?

In the last chapter I advanced a conception of active knowledge, and I will go on to propose revised notions of truth and reality, and then a new perspective on scientific realism. But you may fail to see the motivation for engaging in such an enterprise, if you are quite content with widely accepted positions on these issues rooted in the propositional view of knowledge. Many philosophers, scientists and others take the following set of views as common sense: scientific knowledge consists in having true theories; a true theory is one that corresponds to the world; and the world is just the way it is, regardless of how and what we think about it. How could anyone not agree with all that? And as long as people are content with this standard realist outlook, they will see no point in messing with the well-established fundamental notions for an uncertain payoff. Therefore, before I go on with my positive story I must explain why I think that *all is not well* with this philosophical common sense.

The central notion in the common-sense view is the *correspondence* between our theories and the world. This idea is, of course, closely related to the *correspondence theory of truth*, which holds that 'truth is correspondence to, or with, a fact' (David 2016). Let us see the *Wikipedia* definition of that theory, since we are talking about what is taken as common sense: 'the truth or falsity of a statement is determined only by how it relates to the world and whether it accurately describes (i.e., corresponds with) that world'. The core intuition here is widely shared: truth is a matter of whether our idea (or theory, belief or hypothesis) matches the facts of the matter; we say that the theory is true by virtue of its correspondence to reality.

Many philosophers put the idea of theory–world correspondence at the heart of the doctrine of scientific realism, which is based on the intuition that our best scientific theories give really true descriptions of the world. Of the myriad of definitions of scientific realism in the literature, I find those by Rein Vihalemm (2012, esp. p. 10) and Ilkka Niiniluoto (1999; 2014, esp. p. 159) the most insightful and clear. I follow them in characterizing standard scientific realism[1] as adherence to the following basic tenets:[2]

(1) There is mind-independent reality.
(2) Truth consists in a correspondence between statements (or theories) and reality.
(3) It is possible to obtain knowledge about mind-independent reality.
(4) Attaining truth about reality, including its unobservable aspects, is an essential aim of science.
(5) Modern science has been largely successful in this aim.[3]

Philosophers of science have debated the realism question endlessly, with a particular focus on Tenet 5. But often the first four tenets are assumed to be settled even by many who argue against Tenet 5. Then the debate becomes *unphilosophical*, in the sense of being conducted on the basis of an unquestioning acceptance of certain fundamental assumptions. Traditionally, instrumentalists and positivists in their anti-metaphysical attitude regarded Tenets 1 to 4 with suspicion, but these positions have few adherents these days. Bas van Fraassen's fresh questioning of Tenet 4 was already forty years ago, and his 'constructive empiricism' is respected but has few adherents. Especially Tenets 1 and 2 have tended to escape scrutiny. These two tenets make up what I will call **correspondence realism**: the truth of theories consists in their correspondence to the mind-independent world.

This chapter is devoted to a critique of correspondence realism. As it often happens in philosophy, on closer inspection this common-sense idea turns out to be more like nonsense – not so much wrong as 'not even

---

[1] 'Standard scientific realism' is a phrase that I have adopted from Vihalemm. It is a well-enough recognized notion to be a category in PhilPapers (https://philpapers.org/browse/standard-scientific-realism, last accessed 2 October 2021).
[2] I hasten to add that Niiniluoto and Vihalemm themselves do not fully subscribe to standard scientific realism. I have phrased these tenets in my own simplified way, hopefully without doing violence to the original formulations.
[3] Tenet 5 is not explicitly included in Vihalemm's list. Niiniluoto's formulation of it is as follows: 'the best and deepest part of such knowledge is provided by empirically testable scientific theories which postulate non-observable entities to explain observable phenomena'.

wrong', because it is not clear that it means anything well defined. For a statement to qualify as true by correspondence to the world, that correspondence must be some sort of a match demonstrated through comparison. But how are we to compare a statement with the world, when the world is not composed of statements? Otto Neurath pointed this out very plainly long ago, at the height of his activities as one of the leaders of the Vienna Circle: '*Statements are compared with statements*, not with "experiences", not with a "world" nor with anything else' (Neurath [1931] 1983, p. 66; emphasis original). So we seem to have a basic category-mistake in the notion of statement–world correspondence (or theory–world correspondence).[4] Unless this problem is dealt with, no sense can be made of the notion that we should go and seek such correspondence.

Wittgenstein in his early work proposed the striking solution of declaring that the world itself consisted of facts, so that it could correspond to statements. That was the very opening of his *Tractatus Logico–Philosophicus* (1922, pp. 30–1): '§1. The world is everything that is the case. §1.1. The world is the totality of facts, not of things.'[5] This is reminiscent of the old idea that the 'book of nature' is written in mathematics so it can match our theories expressed in mathematical terms. I have always found the notion of the 'book' of nature a bit strange (also 'In the beginning there was the Word'). But even those who wouldn't go so far as to say that the world metaphysically consists of facts seem quite content to maintain that there are objective facts *about* the mind-independent world, which can correspond to statements occurring in our theories (or implied by them). Let us examine more carefully this notion of the mind-independent world, to which our statements or theories are supposed to aspire to correspond.

Two Senses of Mind-Independence

Correspondence realists say: there is a real world 'out there' that exists entirely apart from our knowledge of it, about which there are certain facts, and the task of our theories is to match those facts. The notion of mind-independence is central in this line of thinking. For example, Tim Button (2013, pp. 7–8) identifies 'The Independence Principle' as the first principle of 'the Credo of external realism': 'The world is (largely) made up of

---

[4] Neurath's own inclination was to dismiss the talk of the 'world' altogether, as the passage just quoted continues as follows: 'All these meaningless *duplications* belong t o a more or less refined metaphysics and are therefore to be rejected.'

[5] '§1. Die Welt is alles, was der Fall ist. §1.1. Die Welt ist die Gesamtheit der Tatsachen, nicht der Dinge.'

objects that are mind-, language-, and theory-independent.' Button takes external realism to be more or less synonymous with metaphysical realism, 'capital-R' Realism, robust realism, desk-thumping realism, and genuine realism. This position is also said to invoke the 'God's Eye point of view', in Putnam's idiom. (I hasten to add that Button himself does not endorse external realism in any straightforward way.)

There is a great deal of confusion created by the notion of 'mind-independence', and I think disambiguating this concept can really spare us a lot of philosophical trouble (and hot air). I propose to distinguish **mind-control** and **mind-framing**, as I will call them. The valid realist intuition (which I fully share) is that by mere thinking we do not decide how reality is. The issue here is mind-control: our minds do not have direct wilful control over any reality outside ourselves; reality does not obey our commands. The urgency in the correspondence realist insistence on the mind-independence of reality has been in their argument against idealists, relativists, social constructivists and social constructionists, or some crude caricatures of such enemies at any rate. We want to avoid what Matthew Brown (2013) has discussed as the problem of wishful thinking. If something is not mind-controlled, first of all its continuing existence does not depend on what we think (or whether we even exist); this is what is called 'ontological' mind-independence by Niiniluoto (1999, pp. 26–7), and also by Sam Page (2006, p. 322), who distinguishes as many as four senses of mind-independence. Lack of mind-control also involves what Page (ibid., p. 325) calls 'structural' mind-independence, which something has 'if it has structure independent of how we say it is structured'.[6] As I will explain in more detail in Chapter 3, the absence of mind-control is at the core of our notion of what real entities are.

Mind-framing is about our conceptualization of entities, and that is a wholly different issue from mind-control. Whenever we designate an entity by a concept, the entity is framed by the concept, which resides in our minds (or our languages). My view is that real entities are mind-framed even though most of them are not mind-controlled. Real entities are mind-framed simply because *all* entities are mind-framed. All that we can ever think or talk about are mind-framed entities. As Putnam (1977, p. 496) once put it in an impactful tautology: 'The world is not describable

---

[6] Various aspects of reality are also causally dependent on us, in the sense that we change them by our actions, or even create them – this is what Page (ibid., p. 323) calls 'causal' dependence. But since we exert these causal influences through bodily action rather than by mere thinking, the objects in question are not directly *mind*-controlled.

independently of our description.' That is to say, any statement we make, any proposition we entertain, has to be expressed in terms of *some* concepts. To think otherwise would be like imagining that we can speak without using any words. All that we can ever talk or think about are conceptualized entities (that is, entities specified by some concepts), not some neat pre-packaged elements that fall out of the universe without any human touch or thinking. The very proposition 'X is real' or 'X exists' cannot be expressed without using the concept X. Concepts are creations of epistemic agents, or at least they are bound to epistemic agents. (I will say more about mind-framing in Section 3.2.)

With the notion of 'mind-independence' thus disambiguated, there is no reason why anyone should be alarmed by the fact that real entities are mind-framed, because that is not linked at all with mind-control. It is not being (entirely) mind-controlled that is the chief hallmark of something real, as opposed to something imaginary or made up. Even my own limbs, which are mind-controlled in their motion, are not mind-controlled in their other properties and dispositions. This is what is behind the widespread and sound idea that reality offers 'resistance' to our wishes and preconceptions. According to Page's interpretation, the 'individuative' mind-independence of the world is the key thing that was denied by some prominent late twentieth-century critics of scientific realism such as Nelson Goodman, Richard Rorty and (one phase of) Putnam, which means that they believed reality to be mind-framed. But as long as they did not believe reality to be mind-controlled, the excited realist reactions against them were misdirected.

The ideas that I am presenting here have been expressed repeatedly in the history of philosophy, usually in the idiom of 'conceptual schemes' or 'linguistic frameworks' or such, by a wide variety of philosophers starting with the neo-Kantians, reaching through to Rudolf Carnap and Thomas Kuhn. These thinkers have expressed the view that entities amenable to cognition can only be meaningfully conceived within one conceptual scheme or another: therefore they are all mind-framed, in different ways. Many people have extended this thought to the whole of human discourse, as with Carnap in 'Empiricism, Semantics, and Ontology' (1950), or to the entire universe, as with Kuhn ([1962] 1970, p. 111) waxing lyrical about scientists living in a different world after a scientific revolution.

The usual notion of 'mind-independence' indicates the lack of both mind-control and mind-framing, and therefore it is a very blunt tool of thinking. The problem is that those who rightly wish to rule out mind-control feel compelled to deny the presence of mind-framing, because they feel that they have to defend the mind-independence of reality at all costs. Worse yet, using the blunt notion of mind-independence allows correspondence realists to *exploit* the fear of mind-control in order to dismiss mind-framing. The conflation of mind-framing and mind-control is at the heart of many fallacious arguments, usually designed to shore up correspondence realist intuitions. One particular version of this argument is what Niiniluoto (2014, p. 168; also 1999, p. 40) calls the 'argument from the past' for 'ontological realism':

> whatever existed before t_0 [the time at which humans appeared in the evolutionary process] must be ontologically independent of human mentality and human cultural constructs such as concepts. This refutes all philosophical doctrines ... which literally claim that nothing exists independently of the human mind or that all objects are our constructions.

What is clear, of course, is that the pre-human things are not controlled by the human mind – that is, we cannot make them do what we want (including coming into being). But that is no argument against the recognition that past entities must be framed by our current conceptions if we are to consider them at all.

The ambiguous notion of mind-independence is used very often in debates on realism, by authors on all sides. The *Stanford Encyclopedia of Philosophy* entry on 'Realism' by Alexander Miller (2016, sec. 6) stresses 'the independence dimension of realism, the claim that the objects distinctive of the [subject] area exist, or that the properties distinctive of the area are instantiated, independently of anyone's beliefs, linguistic practices, conceptual schemes, and so on'. As background to all the arguments to be made in the rest of this book, it is very important that we clearly distinguish mind-control and mind-framing. Whenever we encounter the unqualified notion of mind-independence in a philosophical argument, we should demand to know what exactly is meant by it.

The Fallacy of Pre-figuration

Having disambiguated the notion of mind-independence, I can now return to my critique of correspondence realism in a more productive way. It is a deep-seated intuition, that there has to be *something like correspondence* – because the world must have a certain shape to it, and accurate theories must in some way express that objective shape of the world. But what is not so innocent is the very notion that the world *has* a certain 'shape' without being conceived by the mind. This is the crux of the matter in the debate about correspondence realism. Now, the realist intuition, which most people share, is that *surely* there are things that exist 'out there', which are as they are, regardless of how we think about them, or whether we think about them at all. But this is the great illusion to be dispelled. I am going to call it the **fallacy of pre-figuration**: assuming that reality has well-defined parts and properties that exist independently of all conceptualization. It amounts to assuming that the objects of our knowledge are not mind-framed, yet can be the objects of our discourse, which are necessarily mind-framed.

I am certainly not the first person to note this problem, though the terminology of 'pre-figuration' is my own. It is expressed in the irony noted by Torretti, quoted in the Introduction; Kitcher (2001) and Teller (2001) have made the point over many years, and Rorty (1982) before them. Putnam argued that there was no 'ready-made world', and that there was no merit in claiming to take a God's Eye point of view (Putnam 1981, p. 49; 1982). Raimo Tuomela (1985, ch. 3) discussed it critically as the 'myth of the given', a phrase originating from Wilfrid Sellars: 'There is an ontologically given, categorically ready-made real world'.[7] Long before that, Ernst Cassirer ([1910] 1953) opened *Substance and Function* by attacking the idea that concepts were formed through abstracting from mind-independent descriptions of actual objects.[8] More than anyone else, Dewey in his long philosophical career returned to this issue time and again, perhaps because people continued to miss his point. Sometimes he called it the problem of 'hypostatization' (see Fesmire 2015, pp. 64ff.), sometimes 'the philosophic fallacy', 'which converts the results of knowledge into antecedent realities', and sometimes the 'retrospective fallacy'. I am giving a new name to the problem so that I can connect with these other positions without committing to all the other implications and connotations carried by them.

Pre-figuration is a fallacy because any entity that we can speak or think about has to be specified in terms of some concepts. This is a fundamental point that was perhaps first articulated by Kant: there is no escape from mind-framing (to put it in my own terms). All that we can think or speak about are *phenomena* in the Kantian sense, and the underlying mysterious existence that somehow produces the phenomena in our lived world is the realm of *noumena*. In related Kantian terminology, phenomena and noumena can be understood roughly as 'appearances' and 'things-in-

---

[7] Tuomela also gives two other formulations of the myth. On Sellars see DeVries (2005, pp. 114–16).
[8] I thank Sarah Hijmans for alerting me to Cassirer's argument there.

themselves'.[9] Phenomena, framed by our concepts, are what we engage with in life. To return to my own terms: all phenomena are mind-framed.

But why not just leave it at that? Why even have the notion of noumena or things-in-themselves, if we cannot say anything about them anyway? Kant himself explains: 'it also follows naturally from the concept of an appearance in general that something must correspond to it which is not in itself appearance, for appearance can be nothing for itself and outside of our kind of representation' (Kant [1787] 1998, p. 348). So he thought that there were things-in-themselves, or noumena, which are objects of 'intellectual' (or, 'non-sensible') intuition (ibid., p. 347). What exactly Kant meant by all this is complex and subtle, and my own preference is to avoid the whole debate about the nature of noumena. In order to do that, we need to come away from the notion of 'appearance', which immediately invokes the idea of 'reality' lying behind the appearance as Kant points out. Rather, I think we can take 'phenomena' in a fairly colloquial sense, as what we actually experience, observe, consider, discuss and theorize about. Phenomena are realities, or attributes of realities, as I will discuss further in Chapter 3 when I try to say what we should mean by 'real' and 'reality'. Phenomena are mind-framed but not mind-controlled, and they provide that 'resistance' given by reality. There is no need to postulate some other realm of being aside from, or 'behind', the realm of phenomena. If you really insist, we can say that there are noumena or things-in-themselves, but they should come clearly labelled with the Kantian warning that our concepts cannot be applied to them. And there is great futility in the insistence that the world exists and that it is just how it is, but in a way that cannot be specified. What is operative, instead, is our assumption or knowledge that certain specified things are real – grackles but not ghosts; alpha-rays and beta-rays but not N-rays; electromagnetic fields but not the aether, and so on.

The fallacy of pre-figuration amounts to treating noumena as phenomena, or imagining that things-in-themselves can and should be characterized in terms of our concepts. As I will discuss further in Section 3.1, the fallacy is enhanced by the widespread habit of designating the noumenal realm as 'the world', a term that brings with it clear connotations of the phenomenal realm, the world we live in. If the

---

[9] A brief yet authoritative discussion is given by Nicholas Stang (2018, sec. 6.1), who points out that not just any appearance counts as a phenomenon: 'For instance, if I have a visual after-image or highly disunified visual hallucination, that perception may not represent its object as standing in cause-effect relations, or being an alteration in an absolutely permanent substance. These would be appearances but not phenomena.'

correspondence-schema is meant to be between theory and the noumenal
'world', it is philosophically ungrammatical, because there is nothing
identifiable on the side of 'the world'. Any facts that are capable of
matching with our theories must be articulable in terms of some concepts,
which means that they belong in the phenomenal realm. And if the
presumably existent entities are noumenal and therefore not *specified* in
terms of any concepts, then there can be no articulable factual propositions
concerning them.

Hans Reichenbach expressed this point perfectly long ago, in his discussion of the 'problem of coordination'.[10] If we are thinking about the
coordination between two mathematical structures, the matter is straightforward: 'For example, if two sets of points are given, we establish a
correspondence between them by coördinating to every point of one set
a point of the other set.' In contrast, 'the coördination performed in a
physical proposition is very peculiar' because for the purpose of coordination 'the elements of each set must be *defined*', and that definition is
lacking on the side of 'reality':

> Such definitions are lacking on one side of the coördination dealing with
> the cognition of reality. Although the equations, that is, the conceptual side
> of the coördination, are uniquely defined, the 'real' is not. On the contrary,
> the 'real' is defined by coördinations to the equations. (Reichenbach [1920]
> 1965, pp. 37–8; emphasis original)

Reichenbach's problem of coordination cannot be handled by fallibilism:
the difficulty is not that we do not know for sure how to match up theory
and world correctly, but that things in the 'world' are not even defined
before the coordination is carried out successfully. Put more succinctly, the
problem is that 'we are faced with the strange fact that in the realm of
cognition two sets are coördinated, one of which not only attains its order
through this coördination, but whose elements are *defined by means of this
coördination*' (ibid., p. 40; emphasis original).[11]

Recognizing pre-figuration as a fallacy makes it difficult to talk smoothly
and generally about reality, which is actually not a bad thing. We have to
handle the terminology carefully. As already indicated, I think we should

[10] I thank Raffael Krismer and Milena Ivanova for alerting me to Reichenbach's work in this regard.
What will be more recognizable to others is that Reichenbach provides a central problematic for van
Fraassen's (2008) discussion of scientific representation.

[11] Reichenbach's own solution (ibid., p. 43) is that what makes a coordination correct is its
consistency, with an approving nod to Moritz Schlick's definition of truth as unique
coordination. My sense that this is an insufficient solution should become obvious in Chapters 3
and 4.

use the term 'real' to describe entities that are mind-framed but not fully mind-controlled, as I will discuss further in Chapter 3. If *reality* is mind-framed, then it is perfectly sensible to compare the content of our theories to the shape of reality. Then theory–reality correspondence is something that we can achieve, recognize and test in practice. However, we would lose the claim to be talking about anything that is totally independent of our conceptions – which is fine. But, again, you might ask: if we use the term 'reality' to designate mind-framed things, then what do we call the noumenal Being that is there apart from human mental existence, which must be somehow responsible for the fact that the real entities we conceive are not controlled by our minds? I think it would be best *not* to have a word for that in philosophy! Consider how modern science gets on very well by having no term for God, though that doesn't mean that science denies the existence of God; the point is that it is better not to have a word for something that we have no capacity to talk coherently about.

Overcoming Correspondence-Realist Intuitions

There are, of course, various considerations that make so many people inclined towards correspondence realism. I want to address three broad classes of reasons, and argue that none of them is compelling. They all amount to different forms of the fallacy of pre-figuration.

(1) It is often taken for granted that any empirical statement we make is truth-apt (has a truth-value, i.e., is either true or false); it is true if it corresponds to the state of the world, and false if it does not. Then it seems inescapable that correspondence realism is the only stance we can take about this situation. But when we think like that, we are forgetting that only propositions with determinate meanings can be truth-apt. Meaningless propositions are 'not even wrong'. To give propositions determinate meanings, we need concepts with determinate meanings. As I will argue in Section 2.2, such concepts only exist in the context of specific epistemic activities, which also indicates a high degree of mind-framing, and it is difficult to see how they can be used to formulate propositions that correspond to mind-unframed states of the world. What normally happens at the intuitive level with correspondence realists is that they do populate their picture of reality with mind-framed entities of sufficiently determinate meaning (such as semiconductors and quarks, and cats and mats), but then turn around and declare them to be

mind-independent (free of both mind-control and mind-framing). A vivid and instructive caution against this move is given by Rasmus Winther in a fascinating account of the history and philosophy of map-making and map-thinking, in his *When Maps Become the World* (2020).

(2)   Some correspondence realists argue that the indeterminacy of meaning can be overcome by means of reference. They believe that some types of entities, such as 'natural kinds', are completely mind-independent,[12] and that we *refer* to them by somehow pointing at them with a metaphysical finger. And then, once we have got a referential hold of these entities, we can describe the rest of the world in terms of them. This attempt to avoid mind-framing through reference also fails, as I will argue in Section 2.3. What do we actually do, in order to refer to a chunk of reality? Conceding that the usual method of specifying the recognizable properties of the putative object would involve mind-framing, correspondence realists tend to advocate the ideal of purely extensional reference, simply by identifying the set of objects that fall under the concept in question. But this only pushes the question around without answering it: if we consider how anything is identified as a particular *kind* of object, it quickly becomes obvious that it requires the specification of certain properties.

(3)   There is also a roundabout way in which people can become convinced about correspondence realism: we know that science gives us true theories that correspond to the state of the world; therefore, there must *be* such truth about the world. In other words, if we accept standard scientific realism, then we accept correspondence realism, which is part of standard scientific realism (the first two tenets, in the Vihalemm–Niiniluoto definition of standard scientific realism). But can one argue for standard scientific realism without explicitly arguing for correspondence realism? As I will discuss in Section 2.4, there are major arguments for standard scientific realism that only have their persuasive power due to the faith in science that people have. What happens in these arguments also ends up in the fallacy of pre-figuration, because they start by mentally populating 'the world' with the entities that modern science considers real. For example, one might think that what really exists mind-independently

---

[12] For a nuanced and constructively sceptical view on the notion of natural kind, see the collection edited by Catherine Kendig (2016).

is the distribution of matter (or matter–energy) in space-time, in which people with different conceptual schemes may identify different kinds of objects. But Kant already anticipated such a move. This is why he discussed space and time as human conceptions before anything else in the *Critique of Pure Reason*. Post-Kantian developments have actually enhanced this aspect of Kant's insight even further, by showing that there are different ways in which we can conceive space and time.

'Correspondentitis' and Its Metaphorical Origins

I dare to think that the main message I am trying to convey here is as simple as the emperor of the well-known fairy tale having no clothes on: if the state of the 'world' is something completely free of mind-framing, then it is not something we can ever even think about. As I will say more in Section 2.5, correspondence is a perfectly straightforward and useful notion when we are talking about matching up two things that we can access and describe in the same kind of way. The correspondence between theory and reality makes sense if we mean reality as something conceptualized. But correspondence loses its sense when it is turned into a grandiose notion purporting to capture the relationship between our conception of the world and the 'world itself'.

In my view, the network of notions I am critiquing here constitutes nothing short of an endemic philosophical disease, which we might call 'correspondentitis': an *inflammation* of the notion of correspondence! (Or more politely, I call it 'correspondence realism'.) The aetiology of the ailment:[13] it comes about through a *metaphorical projection* from an operational sort of matching between things that we can actually work with, into the idea of a match between theory and the ultimate reality. Putnam points out that this idea is useless:

> To say that truth is 'correspondence to reality' is not false but *empty*, as long as nothing is said about what the 'correspondence' is. If the 'correspondence' is supposed to be utterly independent of the ways in which we confirm the assertions we make ... then the 'correspondence' is an occult one, and our supposed grasp of it is also occult. (Putnam 1995, pp. 10–11; emphasis original)

---

[13] Knowing the aetiology will hopefully help us devise a cure and stop the disease from returning.

Paraphrasing William James, Putnam opines: 'Truth ... must be such that we can say how it is possible for us to grasp what it is.'

In daily life we do engage in many representational activities in which perfectly meaningful correspondence is sought and gained. If we draw an object that we see, the drawing represents the object and we can check how well it does so; we have direct access to both sides, and we can easily compare them with each other. If we make a map of a terrain, the picture expresses the various geometric relations that we found out by surveying the terrain. If we make a mathematical formula that represents a set of data, we have those data and we can check if the formula represents them accurately. If a witness is testifying in court, in most cases he himself knows very well whether he is lying: he has access to both what he thinks and what he says; his statement is a true or false representation of his thought.

The problem comes when people make a metaphorical projection of such acts of representation onto metaphysical theorizing. The very idea of the 'external' world is a metaphor (outside of what?), created by imagining that there *must be* something that a theory represents, even when we haven't made the theory by actually representing anything. The thinking seems to go: 'We drew a picture, so it must have been a picture *of* something.' After the mental creation of the metaphorical *object*, the rest of the metaphorical structure follows easily: the theory *represents* the imagined object, through some sort of *correspondence* between various aspects of the theory and various imagined properties of the imagined object.

,There are many further metaphors that support correspondence realism by perpetuating the fallacy of pre-figuration. I suggest that we try to get away from all of them. Some of the metaphors posit the existence of pre-figured objects of knowledge that are somehow *hidden from our view*. We say that we are groping around in the dark for bits of reality, latching on to some of them if we are lucky. Or that we are behind a veil (of perception, or whatever), unable to see the objects directly. Or that we have a box containing mystery objects. Even when this metaphor is intended to convey the message that the box can never be opened, as in the brilliant 'box project' by Hardcastle and Slater (2014), it will unfortunately reinforce the notion that there *are* well-defined objects in the box, and the lesson is liable to be mistaken as one about scepticism and human cognitive limitations. Plato's cave allegory, in the way it is commonly understood, is the same kind of metaphor, and naturally leads to the question of whether the poor cave-dweller can get out to see the pre-figured reality outside.[14] Other metaphors do not explicitly posit pre-figured objects of knowledge, but cannot help *implying* their existence. All the familiar visual metaphors fall into this category, including the chief metaphor of perspectivism. Ronald Giere's (2006, ch. 4) perspectivism makes it quite clear that there is one world that is viewed from various perspectives, even though knowledge is always perspectival. The metaphor of looking at something from different angles does presuppose that the common object being viewed is well defined independently of the perspectives. (I will discuss perspectivism more fully in Section 5.3).

The metaphors of the external world, representation and correspondence serve as serious obstacles in the understanding of scientific practices. While the box, the cave and the like are commonly recognized as metaphors, the external world, its representation and the correspondence to it are generally not considered metaphorical. Technically speaking, these are *dead* metaphors – metaphors so ingrained in our way of talking that they are routinely mistaken as literal expressions.[15] For example, as George Lakoff and Mark Johnson (1980) point out, we say that the temperature of a room is going 'up' and that we are feeling 'down' today, as if there were something inherently vertical about warmth or sadness. We may even do ethical harm by taking a metaphor literally, like the people who declared the 'war on drugs' and began to behave as if they were conducting a real war. Dead metaphors provide inexhaustible intuitive fuel for entrenched positions, precisely because they are not recognized as metaphors, and therefore very difficult to eliminate.

Renouncing all metaphors may be psychologically impossible. If so, it would be helpful to offer some *alternative* metaphors, consciously chosen to counter the images of pre-figuration. Some have already been designed, but they do not go far enough. Putnam proposed the 'cookie-cutter metaphor' (1981, p. 52), indicating that the mind-independent world is shapeless like cookie dough, not pre-figured into recognizable and discrete objects; it is only when the cookie-cutters of human-created shapes are applied to the shapeless dough that articulated objects come into being. But this isn't quite right, since a lump of cookie dough does have a well-defined body and particular properties. Niiniluoto's (1999, p. 217) exposition of Putnam makes this worse by rendering the metaphor as 'the world is like a cake', since a cake is even more pre-figured (and less malleable)

---

[14] The disquotation schema can play the same role as Plato's cave picture: 'Snow is white' is true (inside the cave) if and only if SNOW IS WHITE (outside the cave).
[15] See Bowdle and Gentner (2005) on dead metaphors.

than the unbaked dough. The talk of 'noumenal jam' instead of cookie dough is better,[16] but still doesn't quite do the job.

We need a counter-metaphor that provides an image of something that isn't already there but forms as we touch. Picture a supersaturated solution of a chemical that suddenly crystallizes as we touch it – but that is difficult to imagine for those who have not experienced such phenomena. More relatable would be the metaphor of questions and answers: the answer does not pre-exist before the question is asked, unless it is the residue from a previously entertained question. There is certainly no answer to a question that hasn't been formulated. What happens when we are asked about something we haven't thought about? A friend suddenly blurts out: do you love me? First time in Brazil: do you like *jabuticabas*? The answer forms anew, surely based on some inscrutable pre-existing state of my mind, but newly into the framework set by the new question. In scientific research or everyday living, we are the questioners and our interlocutor is nature: we are free to craft the question as we want, but we do not control the answer; still, the question defines the space of possible answers. Or, in Niiniluoto's more correspondence-realist idiom (2014, pp. 162–3): 'We choose the language L, and THE WORLD "chooses" the truth values of L-statements in the world-version W_L.'

For those who insist on the cave metaphor, I suggest that we change the picture of it into a more humble and realistic one. This is my image: if we could step out of Plato's cave, we would only step into another cave. The new cave we emerge into may be a more spacious one, or better in some other way, and accordingly we can even pity those left behind in the old cave, but only in a relative sense. We can never get out of all caves, and there is never going to be that one final escape into the ultimate sunlit reality, but we can keep going, out into yet another cave, and another. 'It's caves all the way out' (in the manner of 'turtles all the way down')! Do we, then, have the picture of an infinitely layered cave-system, which we are forever condemned to navigate through? Not quite – the outer cave is not even there ready-made for us to emerge into. Rather, we spawn the outer cave in the act of emerging from the inner one. We cannot break out of a cave without creating another cave around it. We are the King Midas of caves – everywhere we go turns into the inside of a cave. (By now I have stretched the metaphor so far that I have probably broken it. That is just as well.)

---

[16] Niiniluoto (1999, p. 217; 2014, p. 161) gives this image, referring to Tuomela (1985).

If we reject correspondence realism, how do we move forward? Answering that question is the task of the rest of the book. If you are not a professional philosopher, or if you are already persuaded about the need to move beyond correspondence realism, you may want to move on now to Chapter 3, which will articulate my view on how to think about reality in the absence of correspondence realism. The remainder of the present chapter is devoted to the task of spelling out my critique of correspondence realism more carefully. Sections 2.2–2.4 will take a critical look at three main motivations behind correspondence realism. And then Section 2.5 will discuss how correspondence works in actual practices of representation. All of that material is really only suitable for a professional audience, but the brief summary at the beginning of each section will convey the general idea in a friendlier way.

2.2   Correspondence Realism: Between Metaphysics and Semantics

Certain correspondence realists tend to defend their position by *dissociating* it from scientific realism, trying to secure the notion of truth about mind-independent reality by removing it from the vagaries of actual epistemic activities. But this backfires, because without actual epistemic activities there is no meaning, and with no meaning there is no truth or falsity. That is to say: we can see the difficulty of correspondence realism most clearly by attending to the relation between semantics and truth: no statement can be true-or-false unless it is meaningful, but meanings only exist in our conceptualized practices. This point becomes inescapable especially when we consider how the meaning of a statement might be made sufficiently definite, so that it would have unequivocal truth-value. Whether we are dealing with lofty metaphysics or mundane day-to-day talk, the meaning of our locutions can be pinned down only by a host of well-regulated activities. In Section 2.3 I will examine the attempts to argue that meanings can be secured without relying on activities, via reference. Unless such an attempt works, correspondence realism comes down to the fallacy of pre-figuration, taking it for granted that our statements are meaningful in the absence of what we think and what we do. This is a direction of thought also enhanced by analytic philosophers' tendency to think in 'atomistic' theories of mathematics, namely set theory and mereology, both of which start by postulating pre-figured elements that can be grouped or assembled together without alteration.

Correspondence Realism without Scientific Realism

At least some of the correspondence realists do not subscribe to scientific realism (see Khlentzos 2021, sec. 1). That is an obvious option, when we look back at the definition of standard scientific realism adopted in the last section: remove epistemic ambition and confidence from standard scientific realism (Tenets 3 to 5 in the Vihalemm–Niiniluoto definition given in Section 2.1), and we are basically left with correspondence realism (Tenets 1 and 2). One can maintain one's belief that the mind-independent real world does exist, and that there are objective facts about it, regardless of whether science is capable of finding out those facts. The correspondence ideal of knowledge, and its problems, can actually be seen more clearly in the first instance when we take it away from the question of scientific realism. (In Section 2.4 I will return to the consideration of scientific realism.)

A comment on terminology, before I begin. **Correspondence realism** is a phrase that I made up for this book. The position that I am trying to designate with that phrase is quite akin to what is more commonly known as 'metaphysical realism', which I will sometimes speak of, especially when I am discussing other people's views. Tim Button, who takes 'external realism' as more or less synonymous to metaphysical realism, lists the 'Independence Principle' and the 'Correspondence Principle' as two of the three pillars of external realism. The latter reads: 'Truth involves some sort of correspondence relation between words or thought-signs and external things and set of things' (Button 2013, p. 8, with reference to Putnam).

Metaphysical realism or correspondence realism does *not* require any faith in science. Button (2013, p. 10) identifies the 'Cartesianism Principle' as the last of the three main tenets of external realism: 'Even an ideal theory might be radically false.' That was an implication of metaphysical realism that Putnam had identified, as a problem:

> The most important consequence of metaphysical realism is that *truth* is supposed to be *radically non-epistemic* – we might be 'brains in a vat' and so the theory that is 'ideal' from the point of view of operational utility, inner beauty and elegance, 'plausibility', simplicity, 'conservatism', etc., *might be false*. (Putnam 1977, p. 485; emphases original)

The same issue emerges in Tim Maudlin's attempt to convince Putnam to accept metaphysical realism, which Maudlin defines as follows: 'the thesis that at least some significant parts of language have definite truth conditions such that it is possible for an operationally ideal theory, stated in this part of the language, to be false'. A key step in Maudlin's argument is to separate out epistemology from metaphysics: 'an *operationally* ideal theory, a theory ideal *as far as we can tell*, might actually be false.' In Maudlin's reckoning, there can be a 'metaphysical realist who denies even the mildest form of Scientific Realism' – namely, a radical sceptic (Maudlin 2015, p. 490; emphases original). Maudlin himself subscribes to a mild, fallibilist form of scientific realism, but in Button's characterization external realism is actually strongly committed to radical scepticism, or at least to an allowance of it.

This aspect of metaphysical realism is just the opposite of what many scientific realists seek. Maudlin asks us to imagine that 'the universe as a whole began just six thousand years ago in exactly the physical state that actually obtained six thousand years ago.'[17] In that situation, Maudlin argues, 'reasonable people would conclude that the universe is much, much older than six thousand years' because the 'old universe' theory has 'as much operational utility (no matter how that is understood) as, and more elegance, plausibility, simplicity, etc.', than the 'young universe' theory. So the 'old universe' theory would be operationally ideal, but still false (ibid., p. 491). This is a real possibility, which the denial of metaphysical realism would make it impossible to acknowledge. Maudlin's challenge is based on an insistence on the existence of truth-value for operationally inaccessible propositions; his example is designed precisely so that the truth of the matter is completely inaccessible to any human inquirers. But, the thinking goes, the universe must have started at *some* point in time, so there is a *fact of the matter* about that, whether we can ever know it or not. Maudlin is not saying that he *knows* that the universe was created 6,000 years ago, or not. He is just saying that either it was or it wasn't, and for the metaphysical realist that is the main point – whether science can get at the answer is a secondary issue. This makes metaphysical realism a more *defensible* position than scientific realism, because it avoids the epistemic risk that standard scientific realism incurs by committing itself to Tenets 3–5 in the Vihalemm–Niiniluoto formulation.

The Problem of Unspecified Meaning

If metaphysical realism only asserts that 'the external world exists' then there might be little harm in granting it, and no great excitement in affirming it, either. This is not an issue that well-functioning people have ever worried about, outside some circles in academic philosophy and theology. The

---

[17] This is not a novel idea. As discussed by Martin Gardner (2000) the English naturalist Philip Henry Gosse, in his *Omphalos*, had proposed exactly this idea as a serious scientific theory. That was in 1857, two years before the publication of Darwin's *Origin of Species*.

interesting moment arrives when metaphysical realism goes into full correspondence realism, by claiming that there are some *actual specific statements*
that express facts about the world that are independent of all mind-framing,
statements that are true or false regardless of what we think, regardless of
whether we knowers even exist at all or not.

The best way to register a doubt on this matter is through the
question of meaning. Again, the correspondence-realist instinct is to insist that
any given statement about the world is truth-apt (has a truth-value), regardless
of whether we can know its truth-value, again taking truth to be 'radically non-
epistemic'. And that is the point to which I really object. We cannot just
presume that every statement is truth-apt. As Teller (2017, p. 148; 2021)
emphasizes, a statement with an unspecified meaning does not have a
definite truth-value; if I am talking gibberish, you are not obliged to grant that
what I am saying is either true or false. And there can be no correct answer to
an ill-formed question. Here is Putnam:

> My view is that God himself, if he consented to answer the question, 'Do points
> really exist or are they mere limits?', would say 'I don't know'; not because His
> omniscience is limited, but because there is a limit to how far questions
> make sense. (Putnam 1987, p. 19)

So the burden is on the correspondence realists to show that the words
designating parts of aspects of the external world are meaningful in the
absence of what I have called mind-framing. To the extent that meanings of
words are unspecified, the truth-value of statements that contain those words
will be indeterminate. That much is true, whatever conception of truth one
adopts. Michael Dummett was right when he said (1981, p. 669): 'the theory of
meaning ... is the foundation of all philosophy, and not epistemology as
Descartes misled us into believing.'

But is this a real worry? I believe it is. There are different types of
situation in which what we say is not meaningful enough to be truth-apt.[18]
First of all, there are words that are not involved in enough epistemic activities
that can function as sufficient specifiers of their meaning. Statements containing such words are not truth-apt. Lewis Carroll's *Jabberwocky* enters that
territory intentionally. Is it true that 'All mimsy were the borogoves, and the
mome raths outgrabe'? Or consider statements of the most profound kind,

---

[18] Consonant with my view of active knowledge articulated in Chapter 1, my notion of meaning is
also tied to epistemic activities: a concept or word (or a whole statement) receives meaning from its
employment in coherent activities, which may or may not include the explicit defining of terms.
This may justly be considered a variant of the Wittgensteinian 'meaning as use' idea.

such as 'God is merciful.' Whether this is true would depend entirely on what we mean by 'merciful' (not to mention 'God'), on whether we have got a well-enough crafted concept. The idea of mercy in ordinary human life is simply not suited for understanding what God may or may not be doing. We would need to have enough well-established activities in which we recognize and respond to the mercy of God, and the best theologians are not agreed on how that ought to be done, given the problem of evil and other quandaries – or so it seems, to an ignorant unbeliever like myself. The exact state of theology aside, I think the philosophical point stands: there is no definite truth or falsity to a statement if the words occurring in it do not have full meanings.

There are also imprecise statements, which could be more-or-less true, but not completely true or completely false. If I say 'The modern age began before the twentieth century', that statement is not definitely true or untrue, unless a precise definition of 'the modern age' were to be laid down. Or take a homely example from Niiniluoto (1999, p. 65): the truth of 'It is raining now' will depend on whether we count borderline cases such as drizzle and mist as 'rain'. Or one from Austin (1962, p. 142): 'France is hexagonal' (which I do remember learning in French class) – 'Well, if you like, up to a point.' It could even be that the lack of complete sharpness of meaning is an inevitable feature of any concepts we can employ in any empirical situation. This is the case even in the 'exact sciences' as Teller (2017; 2020, p. 61) reminds us – witness, for example, the messiness of the meaning of 'mass' after the relativity theories and quantum theories have got superimposed upon the Newtonian theory. Teller's overall sense, which I endorse, is that 'science provides us with no statements that are both perfectly precise AND perfectly accurate'; the same lesson can be taken from the work of Mark Wilson (2006). Niiniluoto (1999, p. 65) gives attention to 'semantic theories (such as fuzzy logic) which operate with continuous degrees of truth in the interval [0, 1]'. He says that talking in terms of degrees of truth is 'useful, since many terms of ordinary and even scientific languages are semantically vague. Semantic vagueness should be distinguished from epistemic uncertainty. Degree of truth is, therefore, different from *epistemic probability*'. I will discuss these issues further in Chapter 4.

Whether a statement is semantically well formed may be contingent on the development of science. Physicists used to say things like 'The diameter of an electron is around 10^{-13} metres' and after quantum mechanics we have learned that such statements are quite meaningless. Even more clearly, take a notion like 'a particle with a definite momentum p and a definite momentum x', which was a very meaningful commonplace in Newtonian mechanics, but complete nonsense in quantum mechanics. But if the Heisenberg uncertainty principle were to turn out not to be true, then the phrase just given would turn out to be meaningful again. In real life meanings are not only indefinite but also changeable. This is a great annoyance for any correspondence realists who want to say that pre-figured and fixed truths about the world are expressible in actual statements that we may utter and entertain. The meaning of a word grows and changes with each use it is put to, which can also change the meanings and truth-values of various statements in which it occurs. In an extreme case, the very act of testing a proposition may change its meaning (and therefore its truth-conditions), if the testing procedure includes coming up with a new operational definition for some of the terms in the statement. That often happens in scientific practice.

Candidates for Mind-Unframed Meaningfulness

Correspondence realism requires statements that can express facts that are not mind-framed. Are there any good candidates for such statements, free from the problem of unspecified meaning just discussed above? So far I have brought up some inconvenient examples, but to be fair to the correspondence realists we should instead look to the best kind of case, statements whose truths are definitively established without the vagaries of human epistemic activities.

　　　One idea is that *properly* scientific statements, perhaps those found in the most formalized and mathematical sciences, are free from the indeterminacy of meaning. But even in pure mathematics, where we are at our best with semantic precision, we do have to take some basic concepts or axioms as primitive with no way of specifying or controlling their meanings any further than their mutual definitions in an axiomatic system. Take an elementary example: in Euclidean geometry a point is only defined negatively as something that has no parts or dimensional attributes – what exactly does *that* mean?[19] It receives further, perhaps more significant meaning in its relation to other entities, for example as something that appears when two lines intersect each other, but that brings us into the territory of meaning-as-use; in what further contexts we might recognize the presence of 'points' is not predetermined. In the realm of empirical science, unspecified meanings are only more prevalent. This was a great lesson that a great number of reflective scientists and philosophers took from Einstein's early work on special relativity. What Einstein showed was that there *wasn't* a fact of the matter about

---

[19] Thanks to Seung-Joon Ahn for having got me to think about this one.

whether two distant events are 'simultaneous', unless and until further specifications were made about what simultaneity meant. And if the meaning of simultaneity is specified by means of Einstein's operational procedure laid down in special relativity, statements of distant simultaneity are not truth-apt unless the frame of reference has been specified. There are different ways of giving a precise meaning to a statement, and those ways are devised by epistemic agents. Without such epistemic *work* there is no adequate metaphysics, either – so the neat separation between epistemology and metaphysics is impossible to maintain.

Perhaps it is not useful for metaphysical realists to put too much weight on concepts whose meanings are liable to change in different scientific theories. Maudlin's example mentioned above takes a different approach. Maudlin must suppose that the objects and phenomena mentioned in the statement that 'the universe as a whole began just six thousand years ago in exactly the physical state that actually obtained six thousand years ago' are well defined independently of any particular scientific theories. But are these concepts free from all human conceptualization? *Where* do they acquire their meanings? In this kind of case we are all led to imagine that we know what certain familiar words really mean, but on reflection we don't. I actually have no idea what 'the universe as a whole' really means, but let's leave that aside for now. Consider something more innocuous like the word 'began'. I *think* I know what beginning means because I know what it is for a football game or even a whole geological era to begin. But what does 'beginning' look like *for the whole universe*? How would you produce that scene in the movie? So, one moment there was nothing at all, not even space and time, and then . . . oh wait, if there is no time, then there is no 'moment' there, nor does 'then' mean anything . . . . Or are we imagining the popular view of the Book of Genesis, which assumes that space and time were already there, as was God, before the moment of creation? Maudlin must be thinking like that as he says (2015, p. 492): 'Speakers in the "young universe" and "old universe" scenarios mean the same thing by "There were stars six million years ago," and given what they mean, this perfectly reasonable belief is false in the "young universe" setting.' But to a young-universe *Big Bang theorist*, the phrase 'six million years ago' is meaningless. There are difficulties even if we grant the perspective of God who sat around for a long while and then suddenly brought a lot of matter, out of nowhere, into existence. Never mind the conservation of energy as such, but what does it *mean* to bring something into existence out of nothing? What is the mechanism of such a process? Does it happen completely abruptly at a mathematical point in time, or is it continuous but merely quick? If so, what does the finite-time transition process from non-existence to existence look like? And what is time anyway?[20] Such questions will not go away easily. Maudlin's 'Confessions of a Hardcore, Unsophisticated Metaphysical Realist' ends with a striking admission: 'I have assumed that there is such a thing as an intended interpretation of terms like "star," "year," and "ago" that is precise enough . . ., but I have not even attempted to provide an account of how such an interpretation is fixed' (2015, p. 500).

Would it perhaps be better to leave behind grand ideas, and seek the anchors for metaphysical realism in completely mundane domains of life? The creation of the universe is a uniquely weird event, and Maudlin was perhaps unwise to choose such an example as a vehicle for his argument for metaphysical realism. So let's consider the other end of the unfathomability spectrum, and rest assured in the definite meaningfulness of 'The hen laid that egg more than 30 minutes ago', instead of puzzling over the beginning or creation of time. Let's return to good old 'Snow is white'. No deep metaphysical issues here, right? But the problem of meaning is there, too, albeit in a different form, for everyday concepts bear the 'trail of the human serpent' all too clearly. Does 'snow' exist objectively without mind-framing, so that any given thing is either snow or not, whatever we humans might be thinking or not? Does 'snow' include sleet (and how is sleet different from hail?), and the tiny ice-crystals that form when you breathe when it is really, really cold outside? Snow is a concept defined in the context of earth-bound human life and human activity. This is like Niiniluoto's example of borderline cases of 'rain', but the point is broader than just imprecision; rather, what I would like to emphasize is that the meaning-giving epistemic activities are human undertakings founded on our aims, capacities and conceptions.

By now I think we have exhausted the reasonable options for meaningful and truth-apt mind-unframed statements. The point comes down to the same one expressed in Section 2.1: there are no expressible facts of any kind that are not mind-framed. In the end, if correspondence realists want truth to be 'radically non-epistemic' then the statements that they want to uphold will turn out to be so ill-defined that they will not be truth-apt. In denying this point, correspondence realism on the whole may be just another way of expressing the fallacious idea of pre-figuration, simply assuming that meaningful statements exist apart from all human thinking and acting.

---

[20] Kant strikes again here, even if we don't go so far as Bergson. Time is not a feature of noumenal reality, and 'beginning' could be no more objective than the notion of time itself. Many modern cosmologists and metaphysicians seem to ignore the lesson from Kant's First Antinomy in the *Critique of Pure Reason* (Kant [1787] 1998, A426–9). Once we start asking questions about things like 'the beginning of time', we don't really know *what* we are saying.

Overcoming Set-Theoretic and Mereological Intuitions

Two common habits of philosophical thinking, codified within mathematics as set theory and mereology, drive us towards the fallacy of pre-figuration. These theories formally express the idea that there are immutable *elements* that can be grouped into sets or combined with each other to form larger wholes, without suffering any alterations in the process. We can, of course, build a mathematical theory from any postulates we like, but the problem arises when one presumes that these mathematical theories *of course* apply to reality. It is ironic that this projection of our way of thinking onto the fundamental picture of reality is so often made by those who insist on the complete mind-independence of reality! Set theory and mereology may apply nicely to *some* aspects of reality, but that is a contingent matter. Their unthinking application implies that there *must be* pre-figured objects in reality, and therefore pre-judges the question of correspondence realism.

The degree to which set-theoretic thinking dominates philosophical intuitions can be illustrated by its important appearance even in Putnam's arguments *against* metaphysical realism during his 'internal realism' phase. Even though his purpose was to criticize the idea of the 'ready-made world' made up of 'self-identifying objects' (Putnam 1981, pp. 51–4), this did not stop him from making an abstract illustration of internal realism with the schema of a 'world' consisting of apparently immutable 'objects x, y and z', and then talking about the conceptual latitude in deciding how many objects there are in this world – three according to the Carnapian, and seven according to the Polish logician, who counts the mereological sums of the individuals as real objects (Putnam 1987, pp. 18–19; see critique in Niiniluoto 1999, p. 212). If Putnam's intention was to make a *reductio* of such starting points, that point was lost on many of his critics and followers alike. Metaphysical realists can deal with such examples easily, by retreating to the position that x, y and z *are* real for sure, and the rest is a matter of convention. So Putnam's own set-up relies on the pre-figuration of the elements; fighting metaphysical realism on that basis is a losing game. It is best not to engage in such talk in the first place. This, in my view, is the fundamental problem with all of Putnam's 'model-theoretic' arguments against metaphysical realism.[21]

Similarly, mereology forms a framework for all sorts of pre-figuration, in assuming that a whole consists of parts that magically come together (and stick together) without suffering any alterations and without requiring any sort of 'glue' to hold them together. Again this is easy to think about in

---

[21] For a succinct exposition of Putnam's model-theoretic arguments, see Button (2013), ch. 2.

mathematics, as one can see in a contrived example from Putnam (1977, pp. 489–90), which was also given in the course of his argument *against* metaphysical realism: 'let THE WORLD be a *straight line*', and then let's consider if its parts are points or line segments. In chemistry and physics, atomism (the mereology of the physical realm) has been a seemingly eternal idea. As Chalmers (2009) has shown, atomism was a fruitless metaphysical doctrine until chemists learned how to operationalize the idea of atoms. And it is not the case that chemistry and physics found in nature atoms or elementary particles that are immutable components of all things. As I will discuss further in Section 3.5, according to modern nuclear and high-energy physics, elementary particles are not such immutable building-blocks. Whether the physical world is mereological is an empirical matter, and to pre-judge that issue is to fall into a particularly strong version of the fallacy of pre-figuration.

2.3   Can Reference Save Correspondence Realism?

In this section I give a critique of attempts to avoid the inevitability of the mind-framing of reality by means of a purely extensional notion of reference. This is the idea that at least certain terms in our language can *refer* reliably to pre-figured elements of reality in a totally mind-independent way, even though the full meaning of the terms will be dependent on conceptualizations on our part. This is meant to be in line with extensionalism, 'the general doctrine that no distinction is clear and philosophically significant if it cannot be captured by differences in extensions'.[22] So, if we can state exactly what belongs to the set of objects designated by a term, then we have captured the essence of its meaning. Extensionally referring terms would give us objective anchors that create the possibility of theory–world correspondence. In my view, purely extensional reference cannot be realized because there are no actual methods by which referring can be done without involving intension (the set of qualities or properties associated with a term), at least for generic terms. Underlying the extensionalist notion of reference is the fallacy of pre-figuration again, which *presupposes* the existence of well-defined objects that we can point to, name and group together, without constituting them in any sense in the process.

---

[22] Floyd and Shieh (2001, ch. 9), opening summary of Quine's 'Confessions of a Confirmed Externalist'.

Reference without Mind-Framing?

At the core of the usual realist confidence in the existence of the pre-figured objects is the idea that we can refer to them reliably in a mind-unframed way. Talking about reference rather than the full meaning of a term makes it easier to uphold correspondence realism. Realists can allow all sorts of indefiniteness and fluctuation in the meaning of a term, but still maintain that there is some fixed objective thing that it points to. Presumably it doesn't matter *how* we are pointing to the object, as long as we can tell which object is being pointed to. This was a key part of Putnam's theory of meaning in the standard scientific realist phase of his work (see Hacking 1983, ch. 6 for an insightful critical exposition): the 'stereotype' (commonly acknowledged features) of a term may be diverse and fluctuating, but the reference can remain fixed. And even though our best scientific theories can be quite wrong in some ways, most standard scientific realists maintain that good theories are 'getting at' something real because the key terms occurring in them refer to real things. And the same goes for good, robust everyday descriptions of things around us.

Most correspondence realists assume that reference links the terms in our language (or theory) to real objects without involving any mind-framing. Take, for example, some basic scientific propositions like 'Water is H_2O' and 'Gold is the chemical element with atomic number 79.' Such propositions would appear to provide our basic terms with scientifically sanctioned semantic anchors in reality. And these anchors are regarded as free of mind-framing because they are *extensional*. 'Gold' then just refers to all collections of atoms with atomic number 79, and 'water' refers to all collections of H_2O molecules.

But how is it that our words can get *tied* to the things 'out there', which are not in themselves constituted in linguistic terms? This is where the causal theory of reference comes in, because it allegedly tells us how referring is actually done: someone interacts causally with the object to be named, and sticks a label on it (this is the 'dubbing,' 'christening' or 'reference-fixing' event); subsequently other people learn to use the term in the same way ('reference-borrowing'), and 'underlying their uses of the name are links in a causal chain stretching back to the initial dubbing of the object with that name' (for a clear exposition see Michaelson and Reimer 2019, sec. 2.2). The original dubbing act 'serves to bridge the gap between word and world' (ibid., sec. 3, introduction) – and presumably does so without involving any conceptualization: '(the use of) a name refers to whatever is linked to it in the appropriate way – a way that does not require speakers to associate any identifying descriptive content whatsoever with the name' (ibid., sec. 2.2).

Causal-extensional referential semantics creates strong resistance to any medicine that may be prescribed against correspondentitis. So it needs to be rebuffed before I can do anything effective with the positive ideas I began to develop in Chapter 1. I want to argue that reference without mind-framing is an illusion, which we will be able to recognize if we consider more carefully what actually happens when we give meaning to words. We must attend to *semantic practices*, because we get into misconceptions by ignoring how we actually live and inquire, and imagining something else instead. 'Let us not speak about dubbing in the abstract, but about those events in which glyptodons, caloric, electrons or mesons were named. There is a true story to be told of each … The truths about those events beat philosophical fiction any day' (Hacking 1983, p. 91).

Having Something to Point to

Even with the naming of a single individual person or object, *referring* isn't so free of mind-framing as people might like to imagine. In Hacking's spirit, let's see what really goes on in the giving of names. The most fundamental practice involved here is ostension, or pointing-and-naming. Remember the realist intuition: pre-figured things exist in nature, so that all we have to do is *point* to them (literally or figuratively) and name them. But when my dear friend cries out 'Will you look at *that*!' half the time I don't recognize whatever it is that she is pointing to. Or I think she is pointing to a rabbit, when she is trying to teach me about the embodiments of ancestral spirits (or perhaps she really has internalized the Quinean 'gavagai' idea and she is pointing to rabbit-parts). I can surely point to things that others don't regard as objectively real, and I can only point to things that I can conceptualize as something. If I may riff on Putnam saying that meanings 'just ain't in the head', I would say pointing just ain't in the finger. If any sensible pointing is to happen, what is being pointed to has to be recognized as a distinct object, both by the pointer and by the person receiving the gesture. This is not a trivial process, and it is only because of the presumption of pre-figuration that people imagine that it is.

Contrary to the typical philosophical thinking going back to Saul Kripke, by the time we have a person or thing that we can point to and name, the most important and difficult part of the *work* of referring has already been done. The act of *dubbing*, by itself, has little more than ritualistic significance. The word–object link can be made straightforwardly only if we can identify the object wordlessly. An emperor penguin knows er mate very well as an individual, well enough to pick em out from the huge flat icy field with thousands of emperor penguins that all look identical to us. Sticking a linguistic tag ('Hilary' or something) on this important person is not the essential thing about reference (and not a step that is necessary for penguins to do what they do, as far as we know).[23] By *naming* we do not 'latch on to the world'. If we are able to latch on to anything, then the thing we latch on to has already been prepared somehow, out of the previously unconceptualized world. The identification of well-defined objects in space-and-time is already a great achievement, as Jean Piaget's ([1937] 1954) study of the cognition of object constancy in young children taught us long ago.[24] Discreteness and persistence have to be presupposed, and good enough bearers of those qualities need to be identified in our stream of perceptions.

Now, if it is difficult to secure the reference of proper names without involving mind-framing, for other terms it is downright impossible. But these 'other' terms include *most* concepts that we use in life, including terms that designate kinds of objects, or properties of objects. They include the vast majority of terms we use in scientific theories and observations; proper names only play important roles in a few scientific fields like observational astronomy, ethology and certain parts of the earth sciences. Consider what would be involved in a reference-fixing event for a kind-term. Some primordial interaction in which someone grabbed hold of some gold and pronounced it 'gold'? For that to work, we first have to know how to recognize gold as something distinctive (though not yet named 'gold') when we come across it – and how to recognize it as the same thing when we meet it again, and also how to recognize other samples of gold as the same kind of thing. This is the same as with the case of the emperor penguins recognizing their spouses, but more difficult because now we are asking for the recognition of identity between different individuals, not just the identity of one individual across time and in various circumstances. If we want to point to gold and ask meaningfully if it has atomic weight 79, we first have to know that we have got some 'it' that is distinctive from other stuff. Otherwise the question of what we should call the thing would not even arise.

· To summarize: the common-sense view of ostension assumes that pre-figured objects and object-types already exist, lying around out there to be pointed to. This is, of course, often the actual situation, because we are pointing to objects that have already been successfully conceptualized. But pre-figuration will not do as the basis of a *fundamental* theory of reference that

---

[23] Knowledge-by-acquaintance comes back to bite us at the foundation, after so much express neglect in epistemology.

[24] This lesson stands, even if Piaget was wrong about all sorts of other things. For a more up-to-date view, see Gómez (2004).

gives an account of how those objects are conceptualized in the first place. Here we may usefully hark back to W. V. O. Quine's 'ontological (principle of) relativity' (1969, p. 48; emphasis original): 'reference *is* nonsense except relative to a coordinate system'.

How Sense Underlies Reference

Let us consider the actual processes of reference-fixing further. Think back to Gottlob Frege's ([1892] 1948) classic distinction between 'sense' and 'reference', *Sinn* and *Bedeutung*. Frege's basic picture was that by understanding the sense of a term we fix our attention to the object being referred to; the intension of a term directs us to its extension (reference). As Putnam sums it up: '*reference is given through sense, and sense is given through verification-procedures and not through truth-conditions*' (Putnam 1980b, p. 479; emphasis original). This means that the different dimensions of Putnam's meaning 'vector' are not independent from each other. Specifically, the stereotype is what we use to get to the reference. Take the example of 'gold' again. The individuation and recognition of gold can only be done in a descriptive way (i.e., intensionally) – through its various observable properties, such as its colour, density, ductility, electrical conductivity and resistance to oxidization.

But can't we get directly at the extensions somehow? The Kripke-type view that I am trying to get away from holds that getting at reference through sense is not essential, common as it may be in practice. According to that view, what matters is the correct fixing of reference, not *how* we manage to do it. The memorable example Frege used, of 'evening star' and 'morning star' both being designations for Venus, is deeply misleading. (And I think that the choice of this strongly misleading intuitive example was an important part of Frege's genius! But we must resist that genius.) Giving a proper name to an individual piece of rock is a very unusual kind of thing to do in science. Even for proper names, it is difficult to remove intensions entirely from the picture. Even something like 'Venus' is not an unconceptualized noumenal entity. Venus as a *planet* is already a conceptualized entity. Even Venus as a piece of rock is. Even Venus as a point-like thing with a space-time trajectory is. Again, anything we can talk about at all is mind-framed.

When it comes to terms that are not proper names, it is no use trying to go extensional by claiming that the term 'gold' simply refers to the collection of all instances of gold, unless we can say how to collect up the gold (mentally if not physically) without relying on intensional properties. Recognizing similarity and persistence without intensional descriptions will be a non-starter. As mentioned above, appealing to essential properties such as having the atomic number 79 is no help in this regard, because such properties are also intensional. It may be thought that physical properties can be *rendered extensional* by conceiving them in terms of composition. 'Water is H_2O' will do, and 'having the atomic number 79' should also be understood in this way, as 'containing 79 protons'. But this only pushes the problem one step further. I ask you what the term 'H_2O' refers to. You say its referent is simply a molecule made up of two hydrogen atoms and one oxygen atom. I ask you what 'hydrogen atom' refers to. A composite of a proton and an electron, of course. But what do 'proton' and 'electron' refer to? Eventually, when we have to designate a bottom-level object like an elementary particle, we have to resort to *descriptions*, whether they be fairly phenomenological (specifying its rest mass and electric charge), or very theoretical as in the discourse of quantum field theory.

When it comes to highly theoretical terms, the hope of purely extensional reference fades even further. Theoretical terms do refer to things, but they will not be able to do so without some intricate steps of reasoning through which the theoretical meanings are interpreted in more operational ways, which is squarely in the realm of intensions. To go back to gold for a moment: if we want to *use* definitions like 'Gold is the chemical element with atomic number 79', we have to work out how reference is fixed for 'chemical element' and 'atomic number' (and '79'), and that is not a trivial matter. Attempts have been made to hold on to the causal theory of reference in some way for theoretical terms. The latest notable work in this direction is by Carl Hoefer and Genoveva Martí, who advocate a version of the 'causal-historical' theory of reference. They point out that it is immaterial how a term is initially introduced ('by baptism, ostension of paradigmatic samples, or by description'); what matters is that 'the capacity to refer is passed on, and maintained through the subsequent chain of users of the term, so that even if theories later change ... users of the term are still talking about the same things' (Hoefer and Martí 2020, p. 13). What is causal in their account is not the introduction of the term, but this passing-on of reference: 'The causation involved concerns reference-transmission, i.e., how reference in later uses depends on the existence of a causal chain of reference-borrowing' (ibid., p. 15). But how is that reference-borrowing actually done? They correctly point out that 'bestowing a name' is not a single act but a whole process, and that the successful passing-on of reference requires 'an established referential practice' (ibid., pp. 16–17). I think that is quite correct, and I see it actually as a pragmatist move. But such referential practices will inevitably involve intensional aspects, and fail to realize the vision of purely extensional reference.

The situation was already similar with the account of the causal theory of reference by Kyle Stanford and Philip Kitcher, even though for them the causal dimension is more in the constitution of objects than in the transmission of reference. Stanford and Kitcher acknowledge the complexity of the referential situation, including the inevitable presence of descriptive elements. Their account is in the end based on faith in science, as are other cognate accounts. They rightly point out: 'term introducers make stabs in the dark'. That is, 'they see some properties regularly associated, and *conjecture* that there's some underlying property (or "inner structure") that figures as a common constituent of the total causes of each of the properties' (Stanford and Kitcher 2000, p. 114). If that is indeed the story, then there would be stability in the causal structure of the world. But that wishful story does not say anything about how reference is actually passed down. Something along the same lines can be said about Stathis Psillos's (1999, ch. 12) 'causal-descriptive' theory of reference, even though Hoefer and Martí use it as their foil. In Psillos's account the 'causal' seems to be quite an idle wheel. The key causal properties that allegedly make the object what it is do not seem to play any obvious role in reference-fixing. In short, in all these attempts, what exactly is 'causal' goes murky, and the goal of purely extensional reference is not achieved, since descriptions are needed at every stage of the diachronic process of extension-preservation.

The Plasticity of Reference

Much of the realist confidence about referential semantics rests on the assumption that we can *fix* reference. The most extreme version of this is the notion of 'rigid designators', made popular by Kripke. A rigid designator 'designates the same object in all possible worlds in which that object exists and never designates anything else' (LaPorte 2018, opening sentence). Even those realists who do not want to go into possible-worlds talk would want to insist that at least in this world a term that refers to something real should always refer to the same thing in all circumstances. Therefore correspondence realists are quite disturbed by any prospect of radical scientific change in which key terms seem to change their references. And their opponents knew where to hit them: this is why Larry Laudan's (1981) anti-realist 'pessimistic induction' argument focused on whether terms in successful theories refer, and Kuhn tried to frame incommensurability in terms of cross-cutting extensional taxonomic boundaries (Kuhn [1974] 1977, 309–18; Kuhn [1989] 2000, 76–86).

Meanings of terms in our natural languages or scientific theories do evolve, through a complex iterative process involving shifts in both intension and extension. This is what we witness routinely in the history of science. Elsewhere I have discussed this semantic evolution through the case of some basic chemical terms such as 'acid' and 'element' (Chang 2016a; 2017a). Even in fundamental physics we have such cases. If the question of what 'electron' referred to were simple, Theodore Arabatzis (2006) would not have had to write a whole historical book about it! If we fast-forward to current physics the picture is even messier. Tell me what the term 'electron' really refers to, when it occurs in the following user-friendly statement (being only from *Wikipedia*) of quantum field theory: 'The self-energy of the electron as well as vacuum fluctuations of the electromagnetic field seemed to be infinite.' Does that term really pick out the same thing as what Philip Lenard, J. J. Thomson or H. A. Lorentz referred to by the term 'electron' (or in Thomson's case, 'corpuscle')?

Even the reference of concepts we define at will are subject to empirical contingencies, because whether various phrases containing the concept in question are fully meaningful can be a contingent matter. For example, what the 'centre of the earth' means depends on whether the earth is flat or spherical. If the earth is a flat disk, the reference of 'the country at the centre of the earth' is definite – China (the 'Middle Kingdom'), according to the Chinese. But as the earth turned out to be a sphere, the phrase became meaningless with no possible reference. Questions about the absolute velocity of the earth and other astronomical objects lost their meaningfulness with the advent of relativity theory; more subtly, so did the notion of rigid bodies (see Staley 2008, pp. 278–91). After quantum mechanics it became ungrammatical to speak of a particle with precise values of position *and* momentum.

We can try to hold meaning (including reference) fixed as much as possible, or we can allow it to evolve with the progress of knowledge, according to the demands of empirical inquiry and usage. This is a choice always facing language-users. Even definitions are subject to change, as people's commitments come and go. As Erik Curiel (forthcoming) points out, there are no rigid designators in real science, even in metrology. For instance, we are currently witnessing a deep reform of the metric system, after which the kilogram will no longer be defined by reference to the 'standard kilogram', so that the mass of that carefully engineered and obsessively protected lump of platinum alloy in Paris will no longer be exactly one kilogram by definition (Quinn 2011, ch. 17). In a similar way, the metre stick lost its definitional status long ago. If we must think in terms of possible worlds, then it would be appropriate to acknowledge that human inquirers make different semantic decisions in different possible worlds.

One might try to defend Kripke's notion of rigid designation in the face of such fluctuations of meaning. Kripke (1980, p. 75) says that 'we use the term "one meter" rigidly to designate a certain length', even though 'we fix what length we are designating by an accidental property of that length'. I find it a bizarre notion that something like 'a certain length' exists out there to be referred to, and cannot see how that referring is ever to be actually done. I cannot make any sense of his insistence that even someone who uses something like the standard metre in Paris to define 'one metre' is 'using this definition not to *give the meaning* of what he called the "meter", but to *fix the reference*' (Kripke 1980, p. 55; emphases original). Kripke himself apparently felt the discomfort here but was happy to sweep it aside: 'For such an abstract thing as a unit of length, the notion of reference may be unclear. But let's suppose it's clear enough for the present purposes' (ibid.). Why should we suppose that?

Even if we buy into pre-figuration, assuming the existence of mind-unframed objects 'out there' should not short-circuit our thinking about semantics, which is a question of practice: how do we manage to *give* meaning to our words, and *use* them as effective tools of communication and action? Let us not go into pre-figuration wholesale in the discussion of semantics, even if we do in metaphysics. And most of all, let us not imagine that pre-figuration in semantics can rescue pre-figuration in metaphysics. Niiniluoto's exposition of Alfred Tarski's view on semantics is instructive here. Much attention has been paid to Tarski's discourse about the connection between language and meta-language, but there is also the question of how language and world link up with each other. In Niiniluoto's diagram (1999, p. 56) the reference relation is indicated by arrows pointing from a word in language (English or Finnish) to a pre-figured object in the world, shown as a picture. Niiniluoto (ibid., p. 58) points out that 'Tarski never developed an account of how such language–world relations are established.' Instead, 'Tarski's model-theoretic account of truth presupposes that the object language is interpreted, but it is compatible with various methods of fixing reference.' (If we *start* with that presupposition, we can question-beg all the way to meta-meta land.)

2.4   Faith in Science

Even though this chapter is a critique of correspondence realism, I will briefly address standard scientific realism in this section. This is in order to help eliminate a common roundabout way of arriving at correspondence realism, by the following line of reasoning: science *does* give us the true story about mind-independent reality, so it must be the case that there *is* such truth to be told. I will examine two important arguments in favour of standard scientific realism, and try to show that both of these arguments are weak in themselves, but buttressed by a widespread and strong faith in modern science, which is a symptom of the scientism of our age. First, there is what I have called 'preservative realism', namely the assumption that the lasting elements of scientific theories must embody some truth about reality. The faith in science upholds preservative realism with the trust that if scientists have kept a theory, that *must* have been because it is true; it is not difficult to see the question-begging nature of this move. Second, there is 'the argument from success': the great empirical success of scientific theories can only be explained by assuming that they are really true. Even though there are well-known difficulties with this argument, it appears compelling thanks to appeals to various examples of familiar knowledge, about which we feel that science *must* be in possession of the basic truth.

The Trouble with Starting with the Answers

Continuing with the critique of correspondence realism (Tenets 1 and 2 of standard scientific realism), I now want to address a common *implicit* argument for it. The tenets of standard scientific realism are not entirely independent from each other. Tenet 3 states: 'It is possible to obtain knowledge about mind-independent reality.' If we agree with this, then we agree, a fortiori, that there *is* such a thing as mind-independent reality (Tenet 1). (I prefer not to use the ambiguous designation of mind-independence, but it is unavoidable when I am reporting on typical realist positions.) And to the extent that knowledge involves the possession of truth, the claim that we *can* have knowledge presupposes that there *is* truth to be had about the mind-independent reality. Now, this doesn't strictly imply that the truth should consist in correspondence (Tenet 2), but at least most realists will assume that. In that case, accepting Tenet 3 actually implies the acceptance of correspondence realism (Tenets 1 and 2).

Many people do accept Tenet 3 (that it is possible to have knowledge about reality). People have done so for a variety of reasons, over the course of human history. The word of God, or the direct intuition of a genius, or the power of ordinary reason was thought to give us knowledge about mind-independent reality. Many have lost faith in such alleged sources of knowledge, but meanwhile the faith in science has been growing in the last few centuries (even with some lapses, as we seem to be witnessing currently). It is now taken as educated common sense that science can give us the knowledge of mind-independent reality, and that science constitutes our best hope for having such knowledge. This is why the currently popular flavour of realism is *scientific* realism. Take this from Stathis Psillos (1999, p. 70), a leading advocate of scientific realism who is generally not known for hyperbole: 'what other than our best [scientific] theories should we look to in order to decide what it is reasonable to believe about the world? If our best science is not our best guide to our ontological commitments, then nothing is.'

So if I want to reject correspondence realism, I cannot allow the faith in science to go unchallenged. Most self-respecting scientific realists would object to the designation of 'faith' that I am attaching to their attitude. Rather, they have specific *arguments* that are designed to take us from the demonstrable track-record of science to the conclusion that science must be well placed to give us knowledge of the mind-independent world. I will be dealing with those arguments shortly, but first I want to highlight the backdrop of taken-for-granted assumptions that provide the intuitive fuel for standard scientific realist arguments. If the source of intuitive conviction is not dealt with, no amount of critique heaped on the details of the arguments will have any significant persuasive effect. And as Paul Feyerabend said (1975, p. 25), 'what is the use of an argument that leaves people unmoved?'

Most scientists, philosophers of science and many other people nowadays seem to share a faith in science in much the same way as it was outside the realm of possibility for medieval European scholars not to start from the assumption that the Bible was basically correct, even if many exegetical details could be debated. To get a better sense of the character of thinking going on here, take an emblematic example of knowledge from science: the earth goes around the sun, not the other way around. Most people do not know the actual scientific arguments that led Copernicus and others to believe this.[25] Their conviction cannot be based on their own direct experience, because there is nothing in the direct perception of humans (including astronauts) that shows the earth going around the sun. (This is different from astronauts being able to see the round shape of the earth and, with long enough attention span, the spinning of the earth.) We have such raw conviction about the shape of the solar system only because we've grown up with nice (not-to-scale) diagrams of it. And most people don't think about the

---

[25] For serious discussions of the history of planetary astronomy in relation to the realism question, see Kuhn (1957) and Wray (2018, ch. 1).

fact that modern science has actually rejected the notions of absolute rest and motion (even absolute acceleration, according to general relativity), rendering meaningless the simple-minded question about what goes around what. At best, what cutting-edge science tells us is that if you treated the earth as stationary, the rest of your physics would be enormously and unnecessarily complicated. In any case, the ins and outs of that argument are *not* the reason why so many people are so certain that the earth goes around the sun, any more than most good Christians believe in God because they are masters of intricate academic disputes in theology. For most people, their absolute conviction here is a matter of faith (if not indoctrination) that scientists will have had very good reasons to believe what we've all been taught at school.

Now, one can see why it is psychologically compelling to presume that our own current picture of reality is basically the right picture. If we are asked to suspend our belief in our own picture of reality, then what else can we think with? What Miriam Solomon (2001, ch. 3) calls 'whig realism' may be the only reasonable attitude one can take in practice, just a part of normal epistemic life. General scepticism or agnosticism is an unreasonable demand, which can only be met by the most disciplined of thinkers. But I believe that the task of philosophy is to cultivate such discipline – to think while avoiding blind faith in our present tools of thinking. Recognizing that our currently favoured concepts are not the only possible ones is an essential ingredient in the process of scientific innovation, too, and it is an important task of the philosophy of science to highlight that point, as I have tried to do in the discussion of 'unrestricted inquiry' in Section 1.5.

But many people's instincts rebel against such learned caution, and they circle back to the bald insistence that our current picture of the universe *is* basically the right picture. In more respectable dress, the claim is that we *can* establish the desired theory–world correspondence, and that we *have* done this in some optimal cases. But how is that possible, given that no one has found any principled way of doing so? What we often get is a claim that we *just know* certain things to be objectively true. This move is not an argument, but a *dare*. Some blindingly obvious item of knowledge is presented in front of us, and we are asked if we would dare deny it. If we do, accusations are made that we cannot be serious, that we are insane or wilful and obstinate, and so on. This strategy perhaps goes back at least to the legend of Dr Johnson kicking a rock to refute Berkeley's idealism/immaterialism – 'I refute it *thus*' (recounted in Boswell 1935, vol. I, p. 471). In the modern philosophical tradition, the locus classicus of this move is G. E. Moore (1939) 'proving' the existence of external things by holding up two hands in front of him. There are plenty of scientific examples touted in the same way, too: DNA is a double helix, water is H_2O, the tides are caused by the moon, and petroleum comes from the remains of living organisms. How *could* you doubt such things? But this line of argument, again, begs the questions of why we feel so certain about such propositions, and whether we are justified in that certainty. Much of my previous work (e.g., Chang 2012a) has been devoted to showing that the establishment of such scientific truths is a long and painstaking process, and the outcomes anything but simple and self-evident.

This faith in science is closely related to the fallacy of pre-figuration, but it is not the same, and it is in fact cruder than pre-figuration. Many people take the picture of the world they currently have, and declare that to *be* the shape of ultimate reality, or at least take it as a benchmark against which future theories must be judged. This is a mental habit that seems very difficult to break away from, whether one's world-picture is filled with demons and angels, or the stationary earth (round or flat), or atoms and molecules, or quantum fields and virtual particles. We all walk around with a surprisingly detailed taken-for-granted ontological picture of the world in our heads. We don't all share the same picture by any means, but we all seem very confident about the particular pictures we each happen to hold. If one is a believer in modern science, then the world-picture is whatever one thinks the theories of modern science say. We the scientific faithful believe that what there *really* is in the world is quantum fields and space-time curvature, and perhaps dark matter and superstrings (or, for those who have not paid attention to the recent advances, still the Lego-bricks of protons, neutrons and electrons). And we believe that the universe really started 14 billion years ago with the Big Bang, that the stars really are huge balls of gas powered by nuclear fusion, and so on, and then we say that a theory that corresponds to *that* picture is a true theory.

The whole argument for scientific realism then becomes circular and question-begging: first we decide that the world could only be as it is described by the latest scientific theories, and then of course we will conclude that our scientific theories correspond very well to the world itself! Against this backdrop, finding theory–world correspondence *seems* like a perfectly doable kind of thing. This is related to Bas van Fraassen's comment on how some people deal with Reichenbach's problem of coordination (discussed in Section 2.1):

> the natural temptation in response ... is simply to impose a parallel vocabulary and declare victory. One might say, for example, 'a point in Minkowski space corresponds to a real or physical space-time point, which is a compact convex part of the WORLD with zero measure.' Reichenbach would be quite right to

retort that this parlance is meaningful only *after* the problem of coordination is solved. It cannot be a solution! For this 'solution' begins by thinking of the WORLD as a structure ... described precisely in terms we use in geometry. So it takes for granted that we can represent the WORLD mathematically. (van Fraassen 2008, p. 137; emphasis original)

Van Fraassen is (critically, of course) pointing to a powerful source of intuition that props up various arguments for standard scientific realism: assuming that we have the right picture of the world, which is given by our best scientific theories, we find that our best scientific theories correctly correspond to that picture. This move is so brazen and all-encompassing that it can be difficult to recognize it for what it is. It is really not very different from people getting their preconceived notions confirmed by reading the 'news' created by others who are simply voicing the same preconceived notions. It is viciously circular, and no self-respecting philosopher would invoke it explicitly. But this circular intuition lies at the bottom of many realist arguments, propping them up with much stronger intuitive force than they deserve to have.

I hope to show how the actual steps of inference in realist arguments can be separated out from their intuitive underpinnings, so that they can be evaluated more clearly in terms of their actual merits. That way the arguments will cease to function as elaborate moves dressing up the fundamental intuitive circularity. Scientific realists have two main approaches for actually tackling the epistemological challenge of showing that science does or can attain at least approximate truth about mind-independent reality. I will now review these arguments briefly, and show how weak they are without resting on the faith in science. I hope this will constitute a distinct contribution to the scientific realism debate, though I cannot hope to engage comprehensively with the enormous literature on scientific realism.

Preservationism

The first realist line of argument I want to address is a primarily defensive one, in response to historically grounded scepticism. Here is a pithy statement of the sceptical argument, as phrased by Mary Hesse (1977, p. 271; emphasis original): 'there is the possibility, emphasised by revolutionaries, that *all* our theoretical terms will, in the natural course of scientific development, share the demise of phlogiston'. Given the great fluctuations in what scientists have taken as true, why should we have so much confidence in the truth of what they are currently telling us? This time-honoured argument has derived great force from Kuhn's account of scientific revolutions, followed by Laudan's

(1981) argument that came to be known as the 'pessimistic (meta)induction from the history of science'.[26] The chief realist counter-attack, which I once (somewhat pejoratively) called 'preservative realism' (Chang 2003), begins by pointing out that not everything is in flux. Realism can avoid a direct hit from the fact of scientific change, if *some* theoretical elements can be shown to be preserved through the progress of science, even while drastic changes are made elsewhere. Stable beliefs have a fighting chance of being true. If science does not give us any lasting beliefs, then it does not, a fortiori, give us any *true* lasting beliefs. And if it gives us true beliefs for a moment but they soon disappear, then there is little comfort in that. So it could be said that a fair degree of preservation should be an aspect of any scientific realism worth arguing for.

Going beyond pure defence, standard realists also try to argue that what is preserved *is* worthy of realist credence. 'Selective realism' maintains that certain parts of scientific theories (and only those) deserve realist credence.[27] For example, Kitcher (1993, p. 149) makes a distinction between the (idle) 'presuppositional posits' and the 'working posits' within a theory, which he articulates as the intuition that a statement should play 'a crucial role' in a successful practice in order to be regarded as true (2012, p. 112). This is a nice idea, but the actual work of sorting out parts of theories in this way is difficult to do, so preservation is often used as a proxy for 'working-ness' (being responsible for successes). For example, in Psillos's *divide et impera* (divide and rule) move in response to the pessimistic induction, he wants 'to show that the success of past theories did not depend on what we now believe to be fundamentally flawed theoretical claims'; or positively, 'that the theoretical laws and mechanisms which generated the successes of past theories have been retained in our current scientific image' (Psillos 1999, p. 108). But how is such a thing shown? Even the best philosophers and historians of science, with all the hindsight at their disposal, have great difficulty with this anti-holist task of showing which parts of theories were responsible for success.

So in practice standard scientific realists tend to help themselves to faith in science again, trusting that scientists must *somehow* have figured out how to preserve just those parts of their theories that were really responsible

---

[26] K. Brad Wray (2018, ch. 5) provides an excellent summary and analysis of various arguments that have been advanced along this direction.

[27] Peter Vickers (2017) gives a convenient and insightful summing-up of the state of debate concerning selective realism. A similarly useful critical survey is given by Dean Peters (2012), who speaks of 'partial realism'.

for their successes. With faith, life is easier: one can then just assume that the preserved aspects of scientific theories *are* the success-producing ones. But when we actually take a closer look at key scientific developments, we see that theoretical elements are often preserved for reasons of convenience, aesthetic preference, simple habit or groupthink, rather than their proven role in generating empirical successes. For example, Copernicus retained from Ptolemy the idea that heavenly motions must be composed of uniform circular motion. In biological evolution, too, the survival of a feature doesn't necessarily indicate real contribution to fitness, as Stephen Jay Gould and Richard Lewontin (1979) famously illustrated with the analogy to the decorative spandrels in architecture. On the other hand, theoretical elements actually used in generating success are often rejected by later scientists. In my debate with Psillos about the caloric theory of heat (Chang 2003) I have argued that key assumptions about the nature of caloric really were responsible for the various empirical successes of the caloric theory. Many other cases have been debated in good historical detail more recently (see Vickers 2017, p. 3222, and references therein, including Lyons 2016b). Such problematic cases can be waved aside only we have faith in science. To the faithless, it seems plainly unsafe to trust scientists' judgements about what is the wheat and what is the chaff.

A positive attempt to infer real truth from preservation is particularly popular among structural realists. Consider, for example, John Worrall's (1989) original argument for epistemic structural realism. Inspired by Henri Poincaré, Worrall observes that even through severe scientific change some theoretical elements were preserved, and these tended to be structural ones, usually expressed in mathematical terms. According to Poincaré, our long-lasting best theories get placed beyond question and criticism, becoming 'conventions' that are treated as true by definition. Poincaré was clearly aware that the 'elevation' of a theory to conventional status did not confer empirical truth to it, yet he also hinted that preservation might be linked to truth:

> these equations express relations, and if the equations remain true, it is because the relations preserve their reality. They teach us now, as they did then, that there is such and such a relation between this thing and that; only the something which we then called motion, we now call electric current. But these are merely names of the images we substituted for the real objects which Nature will hide for ever from our eyes. The true relations between these real objects are the only reality we can attain ... (Poincaré quoted in Worrall 1989, p. 118)

Worrall reads the doctrine of epistemic structural realism into this realist moment in Poincaré. In this rendition, preservation comes to serve as a proxy for correspondence-truth.[28]

How is that justified? As Psillos (1999, p. 152) points out, it is an utter non sequitur, and there is no argument there to be found for inferring truth from preservation:

> Worrall needs an argument to take him from the fact that mathematical equations are retained to the conclusion that this retention tells us something about the structure of the world; in particular to the conclusion that the retained mathematical equations represent real relations between otherwise unknown (or, worse, unknowable) physical entities. I am not aware of such an argument in Worrall's (and Poincaré's) writings.

But Worrall's position *seems* reasonable and convincing to many people because the conventionalized structural theories happen to be strongly sanctioned by our faith in current science, upheld as the ideal of scientific knowledge in the twentieth-century gospel according to theoretical physics. There is something subtle going on here: the argument from preservation to truth does not itself rely on the faith in science, but its weakness (or vacantness) is tolerated due to the faith in science. How someone with such a robust sense of healthy scepticism as Worrall has been able to embrace this non-argument has always been a source of puzzlement to me.

Arguments from Success

What van Fraassen (1980, p. 39) has called the 'Ultimate Argument' for scientific realism is based on the idea that the truth of theories gives a good explanation of their empirical successes. This argument for scientific realism from the success of science, or the 'argument from success' for short, is commonly seen as an instance of 'inference to the best explanation' (IBE). The sentiment is: how *else* would we explain the fact that scientific theories are so successful, if it weren't the case that they really get at the truth about mind-independent reality? Two versions of this argument need to be distinguished. (1) One can argue that the success of *science as a whole* implies that there is something about science in general (its methods, social institutions, the epistemic virtues of scientists, etc.) that is conducive to finding correspondence-truth. Putnam's (1975b, p. 73) classic formulation of the argument, for example, operates at this

---

[28] Katherine Brading and Elise Crull (2017) argue that Poincaré's own arguments for structural realism were not so strongly based on preservation (or continuity), but on neo-Kantian notions of what can be general and objective elements of reality.

general level. (2) At the level of individual theories, one can argue that a particular theory that is very successful must be really true. J. J. C. Smart's (1963, p. 39) 'cosmic coincidence' argument should probably be taken at this level of individual theories.[29]

Now, if you take the argument from success purely in the abstract, you will see that it is incredibly weak, at either level. I say 'incredibly' not to be gratuitously insulting, but because I really find it difficult to believe that so many of the most exacting thinkers on the planet can be satisfied with this. The only explanation I can find is a psychological one: all the problems with the argument must appear to them only as wrinkles to be ironed out, rather than deal-breakers for the whole line of thinking. I will just list the main problems with the argument from success here:

(1)   It is actually dubious just how successful even our best scientific theories are.
(2)   There is no clear agreed sense of what we mean by 'success'. (One popular idea is that novel predictions are the best measure of success, but there is no consensus on that, either.)
(3)   There is no clear threshold of success (and its duration) beyond which the demand for explanation should be triggered.
(4)   When we have a successful theory, we are not aware of all possible alternative theories that would do the same job.
(5)   It is not clear what philosophical theory of explanation we are relying on in the argument from success. None of the usual ones (D-N, causal, intentional, etc.) would be suitable for this purpose.
(6)   There is only a very loose logical connection between the success and the correspondence-truth of a theory, due to the need to employ numerous auxiliary hypotheses to apply or test the theory empirically.[30]

Rather than going into the details of those arguments, my focus here is to show how the argument from success is buttressed by the faith in science. At the whole-science level, consider how standard realists might deal with the alternative 'Darwinian' explanation of success given by van Fraassen (1980, p. 40): science is full of successful theories because scientists make many theories and only tend to keep successful ones. This is a plausible explanation

---

[29]   I will not rehearse all the various instances of these lines of argument, or spend time admiring all the refinements that have been made on it; for that, see Psillos's (1999, ch. 4) expert summary, refinement and defence.
[30]   For more on these problems see Chang (2012a, pp. 227–33), and more authoritative discussions in Lyons (2003; 2016a); Wray (2018); Stanford (2018); and Rowbottom (2019).

based on a sufficiently well-understood mechanism (namely, how scientists choose theories). And if we try to make a comparative assessment between van Fraassen's explanation and the standard realist explanation of success, we will realize that it is not even clear what the realist explanation actually *is*. Ask again: why is science prone to producing successful theories? If the story is that scientific theories tend to be successful because science tends to produce true theories, then we have only got ourselves another and even more intractable problem. Why does science tend to produce true theories? Does it? How can we tell, in the absence of a direct method of telling if a scientific theory is really true? (And if we did have such a method, we wouldn't be stuck in the whole argument about scientific realism.) What we come face-to-face with here is a widespread presumption among standard scientific realists that modern science just *does* produce true theories, which is just faith in science.

The argument from success has more of a chance at the level of individual theories, but here, too, arguments in favour of it only look respectable because they are propped up by the faith in science. Standard scientific realists go into battle already believing the basic truth of our current successful-enough scientific theories. This is where the intuitive examples based on the faith in science do their work. We are all exhorted to agree: *surely* we know that the world is made up of discrete particles like electrons and protons and, oh, how successful atomic theories in physics and chemistry are! Likewise, Newton's theory was so successful *obviously* because there is really such a thing as gravity. And if DNA weren't really the double-helix molecule that functions like molecular genetics says it does, how else would you explain all the amazing successes of modern genetic manipulation? Such examples are designed to paralyse the pluralist imagination. In these cases most of us just can't imagine an alternative theory that would be so successful, and we are intellectually bullied by the faithful into agreeing that modern science has basically got the true story, because there can't be any other alternative.

Pluralism can derail the argument from success completely. If there are multiple mutually conflicting theories, all of which are empirically successful to similar degrees, we would not be tempted to explain the success of any one of them by claiming exclusive truth for it. Rather, we would be looking to explain something else entirely, namely how multiple competing theories can all be so successful. Failure of imagination is not a good basis on which to decide a weighty philosophical question. Fortunately, better knowledge of history and current science can help here. Take another look at the intuitively convincing examples just mentioned. The success enjoyed by nineteenth-century atomic chemistry based on the idea of indestructible atoms is not explained by the exclusive truth of that idea; a whole other line of success came from the investigation of atomic structure, which made a joke of the etymology of the word a-tom ('uncuttable'). Even though the success of Newtonian gravitational theory is still undeniable in its familiar domains, for the last century we have known another enormously successful theory, namely, general relativity, that denies the reality of gravity as Newton imagined it. And even though the double-helix structure of the DNA molecule still stands, the original 'central dogma' story of the flow of genetic information now faces serious competition from the more epigenetically oriented accounts. The debates will be much more interesting and instructive if we can start by removing the faith that mainstream science at any given moment basically has *the* true story.

2.5 Real Representations

Now I can start to look more positively again towards an active conception of knowledge, in which the idea of correspondence actually has an important role to play. Correspondence is best taken as a relation that occurs within actual practices of representation, not as a relation that holds between theory and world. To put it in Kantian terms: correspondence that is meaningful in practice connects two items that are both in the phenomenal realm; it does not connect a phenomenal item and a noumenal item. In an actual representational activity we take one entity to stand in for another entity (or create a new entity for that purpose). There are many different types of representing, depending on what kind of entity is being represented and what kind of entity does the representing, and for what purpose. On either side of the relation the entities involved may be material objects, data/information, events/processes, or ideas/symbols/relations. Thinking about what we do in real representations should also help us handle metaphors in a more productive way, so that they can reflect more accurately the realities of practice.

Correspondence in Practice

So far I have cast doubt on the idea that we can ascertain how our words or concepts correspond to mind-unframed bits of reality, or that we can even think meaningfully about such correspondence at all. I now want to turn my attention squarely to actual practices, scientific and quotidian, and note that there *is* a perfectly straightforward notion of correspondence that is operative in practice. As Wittgenstein put it: 'When the words "agreement with reality"

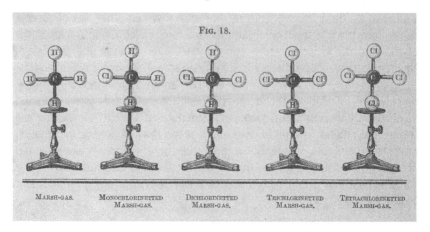

Figure 2.1   Ball-and-stick molecular models by August Hofmann (1865), 425.
Courtesy of Cambridge University Library.

are used, they are not used as a metalogical expression, but as part of a calculus, as part of ordinary language.'[31] In that spirit, let us take the question of correspondence as an *operational* question of epistemology.

Any meaningful correspondence we can actually work with is between two things that reside in the Kantian phenomenal world, not between our conceptions and bits of the noumenal world (to begin with, we can't even say that the noumenal world has 'bits'). In many important cases of claimed correspondence to the noumenal world, what is established is 'ersatz' (fake) correspondence, where it only *seems* that the target of representation is some piece of noumenal reality. But we can rehabilitate these cases as *real phenomenal* correspondence, rather than *fake noumenal* correspondence. For example, a ball-and-stick molecular model may appear to represent real molecules 'out there' (see Figure 2.1). This model does not work as a representation of anything noumenal; what we are actually doing there is making a depiction of a *theoretical* entity, such as the tetravalent carbon (i.e., the carbon atom as conceived to have the capacity to combine with four other atoms or radicals), which is real but fully mind-framed. So these can all be seen as instances of innocent representational activity.[32] Claiming that the model represents Reality will push us right back into the fallacy of pre-figuration. In

[31] I thank Pascal Zambito for this quotation, from an unpublished manuscript (Wittgenstein MS113, 49v, 1931), quoted in Zambito (2019, p. 118).
[32] I believe that my view here is compatible with Adam Toon's (2012, ch. 4) fictionalist yet rule-based account of these molecular models.

that case, what we are doing is first to declare, without justification, that tetravalent carbon *is* the Reality; and then any model that faithfully represents the tetravalency of carbon can be said to correspond to Reality. The question-begging nature of such a move was already pointed out in Section 2.4, in my discussion of the fallacy of pre-figuration backed up by the faith in science.

I do not mean to suggest that we abandon the notion of correspon-dence altogether. Rather, I think we should consider how meaningful corre-spondence obtains in the here-and-now, in this world in which we actually live, by understanding better what goes on in our *representational activities*. Take the slogan from van Fraassen (2008, p. 7): 'It is in the activity of repre-sentation that representations are produced.' I take this to be in the same spirit as Hacking speaking of 'representing' rather than 'representation', taking it as a practical activity just like intervening (despite the famous contrast he drew between the two activities). Van Fraassen explains further: 'Whether or not A represents B, and whether or not it represents the represented item as C, depends largely, and sometimes only, *on the way in which A is being used . . . There is no representation except in the sense that some things are used, made, or taken, to represent some things as thus or so.'* And he is very clear that the use of a representation is to be understood in the context of purposive action: 'our puzzles about what representation is do not disappear unless "use" and its cognates are understood here in the sense in which they presuppose inten-tional activity' (van Fraassen 2008, p. 23, emphases original; also p. 249). As Giere puts it (2004, p. 743): the right schema of representing is '*S* uses *X* to represent *W* for purpose *P*'. In wanting to consider the epistemic activities that constitute representation, I follow the pioneering lead taken by several other philosophers, including Nancy Nersessian (2008), Mary Morgan and Margaret Morrison (1999), and Nancy Cartwright (2019). In the spirit of these great philosophers, let us ask: what it is that we *do* when we represent something?

In very general terms, the activity of representing consists in the creation of a new entity (or the identification of an old one) that expresses specific features of another entity. As Mauricio Suárez notes (2016, p. 453), this is similar to the 'denotation' relation in representation highlighted by R. I. G. Hughes. The activity of representing only makes sense when the object to be represented is already clearly recognized. In slogan form, I would say: we do not *re*present anything that has not already been *presented*.[33] The 'entities' being spoken about here may be of all types, including formal systems. The

---

[33] Katherine Brading and Elaine Landry (2006) use the trope of 'presentation and representation' in their discussion of structuralism. I thank Bob Vos for alerting me to this paper.

representer's intention is to have the representing object ('source'[34]) *stand in* for the represented object ('target'). The linkage of specific features between the represented and the representing objects is the *real* correspondence relation in representation.

As stated already, representation is a purposive activity, carried out in order to facilitate the achievement of certain aims. Representation may help us understand the represented object, by highlighting its key features and suppressing others. It may help us make computations or inferences, as emphasized by Suárez (2004) in his inferential conception of representation. It may help us communicate information, as with graphs and diagrams. It may also inspire us to generate further ideas and hypotheses, or apply pre-existing theoretical notions more broadly. There are also more practical uses, such as using a map to find our way around (see van Fraassen 2008, ch. 3 on map-use). There are many different functions of representation, and a good general theory of representation must be able to accommodate most of the important functions.[35]

What Represents What, and for What?

There are also many different kinds of representation in terms of the kinds of entity that are represented, and the kinds of entity that we use in order to represent them. The variety here may not be a surprise to philosophers specializing in representation, but it will be something new to those immersed in the discussions of mathematized theories representing 'the world'. Material objects, data, ideas, people – just about any types of entities can be targets and sources of representation. As Chris Swoyer puts it (1991, p. 450): 'The diversity of examples suggests that anything can, with sufficient ingenuity and determination, be employed to represent almost anything else, and the uses we make of representations are nearly as varied.' Rather than trying to discuss all the possibilities, I will mention briefly below some source–target combinations that are particularly important in scientific and quotidian life. Recognizing the sheer variety of representational relations will help get us away from the fixed notion that representation is the representation of material 'elements of the world' by theoretical structures.

---

[34] I have always thought that this was a misleading term. Perhaps 'representative' would be better.

[35] The latest work by Roman Frigg and James Nguyen (2020, p. xii) veers away from considering the functions of representation, as they take representation itself as a function (a function of models), while allowing that models have various functions.

We do sometimes represent material objects by means of mathematical structures, and that is the clearest origin of the metaphor of the 'external world' being represented by a theory. But sometimes we represent material objects by other material objects, such as statues and physical models. Somewhere in between physical models and mathematical models as the means (sources) of representation are pictures and photographs. Often the targets of representation are not physical objects, but sets of data or observed facts. A graph representing a set of data points is the simplest example; somewhat more complicated are maps and 'maps' that plot meteorological, geological, or other kinds of data. No less important are representations that express ideas – equations representing the relation between scientific concepts, or schematized human figures representing the idea of gender (and the segregation of sexes in the use of toilets). Traffic lights represent commands to stop or go (or to prepare to do the opposite of what you are now doing). Ideas as targets of representation also include mathematical structures, which are sometimes represented by physical models, and more often by other mathematical structures (as in analytic geometry, which represents algebraic formulas as geometric shapes). People represent other people, in parliaments and local councils all over the world; these representatives routinely also profess to represent certain values, ideologies and interests. In the representational theory of measurement, measurement scales represent the relationships holding between measurement operations. A sequence of events can be represented by a story, or a play. A melody can be represented by a score. In modern art it is not always clear whether or not there is any representation going on. And in numerous realms of life there are also symbolic representations, where a sign stands for something without outwardly resembling the target, which again raises the question of how meaning gets fixed to the sign.

In short, almost any kind of thing can be used to represent any kind of thing, depending on our willingness to try, and the effectiveness with which the attempt serves our purposes. *None* of these exuberantly diverse practices actually involves 'the world' (or even a small parcel of 'the world') as a target of representation. The targets in actual representations are things that are just as accessible as the sources. For example, the four sticks sticking out of the black croquet ball (in the ball-and-stick molecular model shown above in Figure 2.1) represent the quadrivalency of carbon, i.e., our theoretical idea that a carbon atom is the kind of thing that can bond with four other atoms (or groups of atoms). The depiction we have made here is of the theoretical *carbon-atom concept*, not of a thing-in-itself. August Hofmann, who created these ball-and-stick models, knew exactly what the model needed to represent, because he and his colleagues had *made* that theory. The correspondence is easily established and enforced here, because both the model and the target are made by humans. The world, as such, does not enter on either side of the correspondence.

We need to distinguish actual representing from vaguely-and-irresponsibly-aspiring-to-represent. It does not qualify as an act of representing, to imagine an entity and to *say* that our theory perfectly represents it. Real representation requires verifiable and non-vacuous correspondence. And, again, we need to avoid the fallacy of pre-figuration, which will have us imagine that the ball-and-stick picture of molecules *is* the reality, and find satisfaction in how well our theories represent that reality. If we attend to our actual representational activities occurring squarely within the realm of phenomenal rather than noumenal reality, we will find there a practically meaningful and useful notion of correspondence, too. Real correspondence is, for example, between the situations in which we say 'We have water here' and the situations in which we say 'We have H_2O molecules here,' *not* between 'Water is H_2O' and some mythical external-world fact that wordlessly indicates the same. All the objects, operations, depictions and facts on *both* sides of real correspondence relations exist in the phenomenal realm, in the world in which we live – in the same world where there are tables and chairs, that are made of bits of wood, that we lift and move around and stack up.

Progressive Representation in the World, Not Out of It

In Section 2.1 I cautioned against various metaphors expressing the idea of the divide between the world that is knowable to us and the ultimate reality that exists 'out there'. I have argued that such metaphors are deeply misleading. Thinking about actual representation and actual correspondence as I have just sketched out will help us overcome the metaphors of world-representation and world-correspondence, although it will require some disciplined thinking and hard work. While we are engaged in real practices of representation in which correspondence can actually be judged, there is no need to engage in metaphors, and we can simply talk about what we actually do. When we are discussing scientific and other practices, it should not be so difficult to avoid the external-world metaphors. And we should take care to stop the illegitimate transfer of intuitions from those cases to realms in which there are no representations being made and no correspondences to check.

If we must use metaphors, let us do so as conscious masters of the metaphors, not as unaware slaves to them. Let us use them to express specific points, and not take the metaphorical pictures literally in all their aspects. Here is a metaphor that can help us think about real representation, conveying a sense of progress in our knowledge while avoiding the idea of representing 'the world' by something that is other-worldly. It was given in Albert Einstein and Leopold Infeld's popular book on the history of physics (1938, p. 159):

> we could say that creating a new theory is not like destroying an old barn and erecting a skyscraper in its place. It is rather like climbing a mountain, gaining new and wider views, discovering unexpected connections between our starting-point and its rich environment. But the point from which we started out still exists and can be seen, although it appears smaller and forms a tiny part of our broad view gained by the mastery of the obstacles on our adventurous way up.

Compared to something like Plato's cave metaphor, this mountaineering metaphor is more realistic because it gives us a picture of gradual progress, rather than a decisive one-off escape, enlightenment, *nirvana*. Each little vista-point along our way reveals a new view, which also allows a deeper understanding of previous views (quite a lot like my earlier image of emerging into ever-larger caves, but more realistic). And contrary to Einstein's own aspirations expressed in other parts of his work, we should not be seduced by the thought that we are trying to reach *the* top of the mountain, from which we will see *everything*. Somewhere not far away there will be another mountain with an even higher summit, and we must resist the parochialism of regarding our own little summit as *the* top of the world. But *somewhere*, isn't there the highest mountain, the Mt Everest of knowledge? There is, but people flying over the Himalayas know full well that the summit of Everest isn't the highest place there is. To reach the highest place, let's take a rocket, higher and higher ... In outer space we realize that there is no such thing as 'the highest point' in the universe. And at some indefinite point, the idea of a 'high place' turns into the idea of a place *far* from Earth. The very idea of 'the highest place' turns out to be a sadly limited conception, a product of the parochialism of earthlings.

Let us admit humbly that we have no choice but to see things from where we are – whether that be the cave in which we are imprisoned at the moment, or the local summit where we happen to be proudly standing. There is no use in seeking the oxymoronic view-from-nowhere. We pragmatically take as 'reality' the picture that we see from our vantage point. This is a perfectly normal thing to do, and it is also perfectly normal to view previous conceptions as mere approximations of *our* truths, or even as illusions. But let us not fall into the presumption of linear or unidirectional progress. The view from Summit 1 may be denigrated by the view from Summit 2, but the view from Summit 3 may vindicate the view from Summit 1 more than it vindicates the view from Summit 2. Such courses of development do happen in the development of science. The Newtonian physicists derided the ancient belief in a closed spherical universe, but after general relativity, which presumed shape of the universe looks more wrong – finite and spherical, or infinite and 'flat'? Kuhn observed ([1962] 1970, pp. 206–7): 'in some important respects . . . Einstein's general theory of relativity is closer to Aristotle's [theory] than either of them is to Newton's'. The central mistake to avoid is the *apotheosis* of our own inevitably parochial version of reality into the absolute reality, taking our current best picture as the God's Eye view of the world. Our evaluation of other conceptual frameworks can only happen from our own conceptual framework, within which such evaluation can be expressed and justified. Ultimately, we have to find our truths and realities inside *some* cave.

CHAPTER 3

Reality

3.1 Overview

A Pragmatist Take on 'Reality'

I began by considering the nature of knowledge in Chapter 1, articulating my notion of active knowledge and seeing how propositional knowledge functions within it. In Chapter 2 I argued that we should set aside correspondence realism so that the notion of active knowledge can be freely developed. Now returning to a more positive vein, there is an urgent question to be addressed: what is knowledge *about*? What do we have knowledge *of*? When it comes to empirical knowledge, the traditional intuition is that what we (should try to) know about is *reality*. We seek to know facts, which are states of reality. That may be enough of an answer, when it comes to propositional knowledge. But what are the objects of active knowledge? Here, too, we come back to the idea that empirical knowledge is knowledge about reality, since active knowledge is a matter of our ability to engage productively with reality. These vague thoughts need to be articulated more precisely, and without falling back into the notion of reality as completely mind-independent. In order to achieve clarity on all these issues, I will put forward an *operational* conception of reality.

In ordinary English, 'reality' is just a noun form of 'real'. What does it mean in practice when we say that something is real (in the sense of 'existing' or 'actual')? Philosophical debates about reality should connect with the concrete methods by which people reach real-life judgements of what is real. In the discussions concerning realism in metaphysics and the philosophy of science, it is rare that any concrete criteria for reality are laid down. In fact, the version of metaphysical realism advocated by Maudlin and others (see Section 2.2) explicitly removes any inherent connection between metaphysical truth/reality and the operational procedures of scientific and everyday inquiry. Hacking's discussion of 'experimental realism' makes a refreshing departure: 'if you can spray them then they are real' (1983, p. 23). In his view 'reality is parasitic upon representation': or better, 'the concept of reality' follows from the 'practice of representing' (ibid., p. 136). To get a sense of the concrete and operational meaning of reality, we must pay proper attention to our practices, in the spirit of pragmatism as explained in Section 1.6. I also work, to a point, in the spirit of ordinary-language philosophy as advocated by Austin ([1957] 1979, pp. 181–2):

> our common stock of words embodies all the distinctions men have found worth drawing, and the connexions they have found worth making, in the lifetimes of many generations: these surely are likely to be more sound, since they have stood up to the long test of the survival of the fittest, and more subtle, at least in all ordinary and reasonably practical matters, than any that you or I are likely to think up in our arm-chairs of an afternoon – the most favoured alternative method.

Linguistic and scientific practices are surely not infallible guides to philosophizing, but they are sources for plausible insights worth examining. They provide as good a starting point as any in epistemology, with clear implications for metaphysics, too.

In everyday life we do routinely make judgements of what is real. Treating some things as real and others as not is a very important part of how we live. Scientists do the same, though they may not often invoke the terminology of 'reality' in expressing their judgements, preferring to talk about statistical significance and such. *Within* all kinds of concrete practices we know very well how to judge what is real and what is not, arriving at verdicts like 'Ghosts aren't real', 'The Loch Ness Monster isn't real', 'The placebo effect is real', 'Anti-matter is real, and probably dark matter, too', or 'The mesosome, long believed to be a real entity within bacterial cells, turned out to be an artifact of the chemical fixation process used to prepare the cells for electron microscopy.'[1] Scientists sometimes have uncertainties about specific answers to 'real or not' questions in practice (e.g., concerning mesmerism or the Little Ice Age), but they know how to go about deciding the answers, while admitting that their judgements are fallible. In all these situations no one seems to be confused about what being real means,[2] even when there are disagreements about answers to specific questions.

---

[1] For an excellent account of the mesosome episode, see Rasmussen (1993).

[2] A disambiguation of ordinary language is necessary here. The English word 'real' can also mean 'genuine' (the opposite of 'fake' or 'imitation'), 'actual' (as opposed to 'fictional'), or 'exemplary' (as

Operational Coherence and the Meaning of Reality

I propose the following definition as the core of a **coherence theory of reality (real-ness)**: *an entity is real to the extent that there are operationally coherent activities that can be performed by relying significantly on its existence and its properties.* Recall that 'operational coherence' means aim-oriented coordination in an activity, a matter of doing what makes sense to do (see Sections 1.1 and 1.4). And when I say that an activity 'relies' on an entity, the sense of reliance here is not one of metaphysical necessity, but of actual use and need. I think this definition of reality (real-ness) is consonant with many well-established quotidian and scientific uses of the word, and it can also do very useful philosophical work by helping us spell out what realism should mean in the context of empiricism and pragmatism. When we say something like 'The placebo effect is real', we are expressing our judgement that the named thing is *operative* in some processes. Its existence is a difference that makes a difference. When physicists say that positrons are real, or when we say that dogs and cats are real, that means one can do meaningful and effective things with them, like making PET (positron emission tomography) scans, or taking them to the pet shop.

It is important to note that what I am making here is a *semantic* move, in the sense that it concerns the very meaning of the word 'real'.[3] I am *not* proposing operational coherence just as an *indication* of metaphysical reality, which serves as evidence that something is real. Rather, it is about what we *mean* by something being real, and I suggest that there isn't anything else that 'being real' means in an operational sense. (I am proposing a constitutive criterion of real-ness, not an epistemic criterion.) To draw a rough analogy: if you ask me 'How do you tell if you have a headache?' then the answer is 'Of course, I check if my head hurts.' But that doesn't mean that a pain in the head is a *symptom* of a headache; no, a headache *is* a pain felt in the head; that is what the word means. There isn't some Platonic thing called 'HEADACHE' out there, of which the hurt feeling in my head is merely a symptom or manifestation. Likewise, I am

---

in 'a real gentleman'); such meanings are not my immediate concern here (and all these meanings of 'real' would not translate into the same word in all languages). People know not to conflate these different senses of 'real.' This 'Louis Vuitton' handbag isn't real, but yes, it is a real handbag. Harry Potter is a real fictional character, while Parry Hotter isn't real; however, I am not going to try to take Harry to lunch like a real person. (I thank James Tartaglia for prompting me to clarify these distinctions.)

[3] To avoid confusion: this is a very different matter from 'semantic realism', which is a matter of reducing the understanding of a statement to the knowledge of its realist truth-conditions.

proposing that we use the term 'real' to *mean* having the capacity to support coherent activities, in which case it would be misleading to say that an entity is able to support coherent activities *because* it is real. That kind of causal talk would only add an unilluminating and intractable type of metaphysical layer to our thinking.[4]

Many people will be worried that my proposal distorts or perverts the meaning of 'reality' too greatly. That is a legitimate kind of concern, which can be debated further. But first of all let me make sure that the nature of my proposal is conveyed clearly. I am not claiming that the definition of 'real' I am proposing here encompasses every existing usage of the word. Rather, I am proposing that the capacity to support coherent activities is what we *should* mean by 'real', because I think it will be conducive to productive discourse, while being reasonably faithful to enough of the actual usage currently embedded in various practices. So what I am engaged in can be seen as a project of explication as conceived by Carnap, or an attempt at 'conceptual engineering'.[5] The important point to recognize is that we have some choice in what we mean by a term. So, we *could* decide to postulate an unobservable disease-entity called 'head-ache' whose one and only symptom is pain in the head, but we need to ask how that would be a productive move. Any semantic proposal of this sort is to be judged by its fruits in a pragmatist manner, because there is no higher court of appeal.

We should, of course, also address the separate question of how we *know* if something is real. According to my meaning of 'real', we have first-hand knowledge that an entity is real if we (personally) know how to perform some operationally coherent activities that rely on its existence and its properties. If we are aware of some such activities that other people can perform, then we have second-hand knowledge of the reality. If there are coherent activities that someone can perform relying on the entity in question but we are not aware of this, then the entity is real but we don't know that it is real.

---

[4] This is where I diverge even from Hacking's 'entity realism'. As David Resnik (1994) points out, Hacking's argument, contrary to his intentions, collapses into a scientific realist argument from success, taking reality to explain empirical success (see Section 2.4). It makes no great difference here that Hacking is thinking about the success of interventions rather than predictive success. But Hacking's position is less problematic if it is taken as a low-key epistemic thesis. When he says that 'engineering, not theorizing, is the best proof of scientific realism about entities' (1983, p. 274), we should take 'proof' simply in the sense of 'how we know'.

[5] For exemplary work under the banner of conceptual engineering, see Haslanger (2000); Brun (2016); Cappelen (2018); and Dutilh Novaes (2020).

Distinguishing this epistemic condition from the definition of reality should reassure those who fear that my conceptions make the existence of external reality dependent on our subjective knowledge of it. So, yes, a tree does fall in the forest even if no one is there to hear the sound – and all that. Let us not get into an unnecessarily learned discussion of counter-factuals and modalities here. When people say '*of course* the tree fell even though there was no one to hear it', they mean that anyone present would have heard it, that anyone standing on the unfortunate side of that tree would have been crushed, that we can go now and see that the tree has fallen, and so on. You may mean something additional and fancier by the tree 'really falling', but in that case whatever you are asserting there is not *obviously* the case. The real-ness of an entity (or an event) is a matter of whether there are coherent activities it can facilitate, not a matter of whether our current community of people can and will actually perform such activities. So something can indeed be real without us knowing anything about it, and our lack of knowledge does not make anything non-existent.

Still, I expect many people will be uncomfortable with the idea that real-ness depends in *any* way on what we think or do. I think that this worry is at least exaggerated. Consider, for example, what I will call the **inaccessibility argument** for metaphysical realism, which encompasses the 'argument from the past' discussed in Section 2.1. The inaccessibility argument works by pointing to entities that are inaccessible to human inquirers, yet seem undoubtedly real; if there are such entities, then there are real entities even if there could be no human cognition or activity involving them. Surely dinosaurs were real, and so was the asteroid or comet whose impact 66 million years ago wiped them out, but there were no humans then, and therefore no activities performable by anyone involving the dinosaurs or the asteroid. Then do I not have to deny the reality of these entities?

In response, I would first of all point out that it is not the case that we (in the present) are completely lacking in the knowledge of these inaccessible entities. There are coherent *present* activities that we perform by relying on the comet's past existence and properties, such as identifying its traces in geological strata. When we engage in a coherent explanatory activity involving the comet colliding with the earth in the distant past and causing mass extinctions, or a coherent observational and classificatory activity involving dinosaur fossils, then we begin to have a knowledge of the reality of the comet and the dinosaurs. How else does it make sense to say that scientists know that such things are real? And why else *are* we actually so sure that there were dinosaurs?

And so it goes with other kinds of situations of inaccessibility, too, such as entities in a distant corner of the universe that we can never reach, or black holes that we can never directly see (even as astronomers are now busy making images of the 'accretion discs' around them).[6] Hacking's criterion of direct intervention for the knowledge of reality is too restrictive even for pragmatists, and he need not have conceded serious uncertainty about the reality of astronomical objects on the basis of that criterion (see Hacking 1989; Shapere 1993 for this debate). It is not the case that we have to be in the same spatio-temporal location with an entity in order to be engaging with it in coherent activities. Even when I am just looking at an ordinary object not far away, you could quibble and say that the interaction is actually indirect, since it involves photons bouncing off the object and being received by my eyes and causing complicated nerve signals. You could also point out that actually the information conveyed to my mind is about something that existed in the past because the photon leaving the object takes some time to reach my eyes. But nothing of philosophical significance follows from such points. Sure, we cannot really 'go to' the near-past cat that I see walking down the road, but what of it? And this is just a very mild version of the inaccessibility of the dinosaurs of the distant past that I can't take a time machine to go see. The cat has given me observable traces, and so have the dinosaurs. I judge them both to be real, as would the metaphysical realist. No one has scored a point here.

Progressivist Constructivism

A related worry would be that there is a kind of constructivism in my view of real entities. It may seem strange to tie the metaphysical notion of reality to operational coherence, which is based on pragmatic understanding. Doesn't this deprive reality of its mind-independence? The disambiguation of mind-independence I made in Section 2.1 should be helpful here. I take reality as mind-framed, but not mind-controlled. In fact, not being subject to mind-control is an important hallmark of reality; in my previous work I went as far as to say: 'I propose to think of external reality as whatever it is that is not subject to one's own will' (Chang 2012a, p. 220). This accords with an ordinary-language sense of 'reality' as well; the first definition of 'reality' given by my trusty *Collins English Dictionary* (4th edn, 1998) is 'the state of things as they are or appear to be, rather than as one might

---

[6] See Skulberg (2021) for a fascinating account of the history and practice of black-hole imaging.

wish them to be'. Take this as the new realist common sense: all entities are mind-framed, but only a small portion of them are mind-controlled.

Real entities are, at least in some respects, mind-uncontrolled. They do not obey our wishes; even when they do what pleases us, they are not doing *as we please*. Even something we *define* in a purely conventional way cannot be *controlled* as we wish. Take constellations: after we fancifully connect up a certain group of stars as 'Orion' or 'the Big Dipper', we cannot dictate what shape the group will have a million years later when the individual stars will have moved around. That is to say, even though Orion is obviously a mind-framed entity, it is not mind-controlled. The question of reality is *not* a question about unqualified mind-independence, contrary to the metaphysical-realist instinct.

The constructivism inherent in my pragmatist view of reality is not anything that should worry those who seek empirical knowledge, any more than empiricists should fear van Fraassen's *constructive* empiricism (see Boon 2015 for a positive take on epistemological constructivism). It is not that we can create reality just by creating a new concept; however, the presence of a concept is a *prerequisite* for there being any specifiable entity that we can speak or think about. We can make concepts as we like, but whether the entities they specify turn out to be *real* is not up to us. If we do manage to create a new concept that designates an entity with which we can engage in coherent activities, then that is a successful inventive process. This process deserves to be called 'invention' more than 'discovery', but its success is not in our control, and that is just the same in the technological processes of invention. Guglielmo Marconi did not simply conjure up wireless telegraphy in any arbitrary way he fancied; the coherence of his operations constituted a great achievement precisely because it was not guaranteed.

I also want to stress that whatever constructivism that is present in my view of reality does not stand in the way of progress in our knowledge of reality. On the contrary, as I will discuss further in Chapter 5, the pragmatist notion of reality is perfectly suited for encouraging the growth of knowledge, as it helps us elucidate the various forms that epistemic progress can take. We may conceive a new entity, and learn of its reality by successfully crafting some coherent activities on its basis. We may also improve our knowledge further by working to enhance the coherence of these activities. We may also seek to learn if the entity is real in additional domains, by trying to devise coherent activities in those new domains. We may also try to increase our knowledge by coming up with additional activities in an already familiar domain. Generally, our knowledge grows as we learn to engage in more coherent activities and more-coherent activities. There is much more to say about how our learning about reality actually takes place in practice, as I will discuss further in Sections 3.2 and 3.3.

Can There Be Reality from Mental Activities?

There is one other issue that is worth flagging up briefly before I go on. My notion of operational coherence would seem to apply in any setting in life, including the domain of purely mental activities. Does that mean we need to grant reality to imaginary or fictional entities if they can support operationally coherent mental activities? A similar question would arise regarding formal entities postulated in mathematical systems. Are imaginary numbers real because we do carry out many coherent mathematical activities relying on them? I think the answer is 'yes' to all of these questions, and any initial sense of absurdity should dissipate upon more careful consideration.

One quick remedy for the sense of absurdity would be to distinguish different types of reality, going in the direction of a pluralist metaphysical attitude that John Dupré (1993, p. 36) once advocated under the name of 'promiscuous realism': there can be multiple valid taxonomic schemes in the same domain. This thought can be extended further. We can immediately say that an entity has *material* or *physical* reality if it can facilitate coherent activities that treat it as a material or physical entity. The square root of -1, say, is not a material object and does not do physical work in any coherent activities, but it has formal reality in the sense that it plays a crucial role in many very coherent mathematical activities. Likewise, the Excalibur is a real entity in the fictional realm.

But in all these different domains, there is one thing in common: whether an activity we devise turns out to be coherent is not up to us. This is the case even in purely mental activities: the idea of 'a square circle' can be put up verbally, but cannot be executed on paper, or even in visual imagination. Likewise, one can seek the limit of a series, but that will not be a coherent activity if the series is divergent. In this sense, the lack of mind-control underlying the intuitive idea of reality is present in mental activities as well. And so is the lack of pre-determined certainty in the outcome, which is the fundamental characteristic of the empirical domain. Consonant with the spirit of pragmatism as I take it (see Section 1.6), we can recognize that in this sense all activities in life are empirical, and they deal with real entities.

How Not to Talk about Reality

So far I have discussed what it means for an entity to be real ('reality' as real-ness). Now let me address a different sense of 'reality' that is the more usual subject of metaphysical discourse, namely the thing that exists 'out there', the 'world', and so on. I propose to take this concept, too, in the most humble and concrete way possible. Let us take **realities** simply to mean entities that are real. With this meaning, 'reality' is a countable noun, and 'reality' without an article in front of it is not grammatical (unlike 'reality' that means real-ness). It is annoying that 'reality' as the noun form of the adjective 'real' is ambiguous in its meaning, but the two meanings as I propose to take them ('real-ness' and 'something real') are easily consonant with each other. Some other English nouns formed from adjectives also exhibit the same duality: e.g. 'absurdity' meaning both absurd-ness and something that is absurd. It is slightly awkward to say that a sparrow is 'a reality', but if you find it difficult to get used to that usage, you can simply spell it out as 'a real entity' each time. The *Collins English Dictionary* (4th edn, 1998) gives the following definitions for 'reality' (the first of which I quoted earlier):

> 1 the state of things as they are or appear to be, rather than as one might wish them to be. 2 something that is real. 3 the state of being real. 4 *Philosophy*. 4a that which exists, independent of human awareness. 4b the totality of facts as they are independent of human awareness of them.

I am picking up definitions 3 and 2 as the chief meanings, and propose to understand the others on the basis of them.

Now, all this might be quite a let-down to the deep metaphysical mind, and my notion here certainly does not sit so well with the kind of reality that many philosophers like to talk about – namely, 'the world' (or the 'external world'), *the* reality, or 'Reality' with a capital R – the totality of existence.[7] I have difficulty thinking of any coherent activities involving Reality in this sense. Such a grandiose notion only occurs in the kinds of metaphysical, religious or mystical discourse that I don't know how to engage in sensibly, and will not enter into in print. If I innocently stepped into questions like 'Is the external world real?', I would probably never be able to come back out of the rabbit-hole. I think we would do well to avoid thinking in terms of a totalizing kind of Reality, because it is a much-too-

---

[7] C. I. Lewis (1929, ch. 7) employed this device of distinguishing 'Reality' and 'reality' for similar purposes.

lofty construction with no concrete operational purchase. Markus Gabriel ([2013] 2015, p. 12) makes a playful yet significant point when he argues: 'something exists only when it is found in the world ... the world cannot in principle exist because it is not found in the world'. Perhaps more palatable to most philosophers would be Nicholas Rescher's (1980, p. 345) point that terms such as 'the world', 'the universe', 'the true facts' and so on are mere placeholders, vacuous unifiers, designating 'an inherently empty container into which we can put anything and everything'.

It is very difficult to make sense of the claim that 'the world' is real, according to my notion of real-ness based on operational coherence. There isn't anything we can *do* with 'the external world' as a whole. Can we even *talk* meaningfully and usefully about the whole of Reality, or the totality of all realities? Really, what *is* 'the world'? I understand what the 'world' means in phrases like 'world champion' or the 'World Health Organization', as a collection of all the countries or other kinds of human communities (세계 se-gye, in Korean). Similarly it might mean the Earth (지구 ji-gu), our planet, as when we talk about the 'climate map of the world'. But Anglophone philosophers have a habit of saying 'world' to mean something more like the universe (in Korean the universe is 우주 u-ju, and in everyday contexts hardly anyone would think of translating English 'world' into this word). On reflection, it is really not clear what it is that we philosophers have in mind when we talk about 'the world'. If you think that is a straightforward matter, tell me this: does the world include God in it? Even when cosmologists theorize about the universe, they deal with only particular aspects of the universe conceptualized in very specific ways, not just 'all that there is'. I find it difficult to see how we can achieve any sensible aims through the kind of move exemplified by writing 'Ψ' for the quantum wavefunction of the whole universe.

It is not easy to avoid talking about the grand universe-scale notion of reality completely, but I think we should try. When I advanced a doctrine of 'active realism' in a previous publication, I defined it as 'a commitment to maximize our learning from reality', taking 'reality' as that mind-independent something in which we live, which can resist our attempts to deal with it in some particular way that we might prefer (Chang 2012a, p. 220). That was a mistake. As I discussed in Chapter 2, reality in this sense is like the Kantian thing-in-itself, about which we should say nothing; it doesn't make sense to think that we can learn anything expressible about it. As Goodman (1978, p. 4) was at pains to stress, all that we can ever actually deal with are 'versions' of the 'world':

We cannot test a version by comparing it with a world undescribed, undepicted, unperceived … While we may speak of determining what versions are right as 'learning about the world', 'the world' supposedly being that which all right versions describe, all we learn about the world is contained in right versions of it; and while the underlying world, bereft of these, need not be denied to those who love it, it is perhaps on the whole a world well lost.

And even Goodman's way of talking is only a ladder to be kicked away once we have climbed it. 'Versions of the world' is a phrase that inevitably raises the expectation that there *is* such a thing as *the world*, of which we make versions. That invites the accusation that Goodman is denying the reality of the evidently existent 'underlying world'. This is the same problem that I have mentioned in relation to perspectivism in Section 2.1.

Pragmatist Metaphysics

With my proposed conception of reality I am trying to set the scene for what Sami Pihlström (2009) calls **pragmatist metaphysics**. In order to have any kind of reasoning and discourse, we need to conceive identifiable and trackable entities with clear properties. This is the business of metaphysics as I see it. It is commonly thought that pragmatism, like positivism, should avoid metaphysics altogether or somehow dissolve it into something non-metaphysical, but that is not the most productive view to take. I see pragmatist metaphysics as the business of *building good ontologies* that will support coherent activities. In science and other empirical realms of life, the challenging ontological task is to create concepts that specify real entities (or, realities). As I will discuss further in Chapter 5, this is what realists should aspire to do, unless there are particular reasons to take fictionalist or instrumentalist attitudes in some particular situations. Identifying the realities that we should be dealing with is a crucial part of any process of inquiry, and concept-making is not just a matter of thinking up ideas. For concepts intended to specify real things, their uses have to involve coherent arrangements of material and social settings. This task of concept-creation cannot be avoided, and it needs to be done well. Concepts are not simply handed to us by God or through an intuitive access to Platonic heaven, and our inborn instincts are not sufficient to give us good concepts when we need them. Successful concept-building is an 'engineering' process in which we fashion realities, entities that are not subject to mind-control but amenable to our understanding and engagement.

Having a good operational ontology is crucial for any kind of cognitive activity. Many years ago now I was delighted to meet a computer scientist whose business card proudly bore the title of 'Chief Ontologist' for his firm; 'ontology' is a well-established technical concept in his field, and so it should be in every field of research. This point has received a new recognition in the 'data-centric' sciences such as genomics or proteomics with their need to craft the right 'bio-ontologies', as discussed by Sabina Leonelli (2016, p. 26, and *passim*). Without an ontology we cannot say anything intelligible, make any kind of analysis, or engage with nature in any specific and directed way. So there is something else fundamentally right about Hacking's perspective again: realism concerning entities is prior to any realism that we can have concerning the truth of the statements that we make about the entities in question. If the entities that we speak and think about were not real, it would not make any sense to maintain that the statements we make about them are true. It is necessary for us to grapple with ontology if we are to talk about truth.

What I am advocating here is a modest and piecemeal practice of naturalistic metaphysics – not giving a grand view of 'how the world is' arising from abstract reflections removed from experience, but allowing our knowledge of ontology to emerge from well-established practices, in the spirit of Nancy Cartwright's work (1999; 2019). This is much broader than 'naturalism' as it is often meant, which tries to mould metaphysics in line with the propositional content of accepted scientific theories. Rather, we should learn about real things (including the very fact that they are real) by seeing how we can create various operationally coherent activities relying on them. Scientists are seriously engaged in the business of crafting new and better concepts that support ever-multiplying coherent activities. Again, 'salvation is through work', and metaphysicians should pay attention to scientific *work*, with full respect for the ingenuity and sustained effort of scientists. But doing naturalistic metaphysics should not mean a renunciation of philosophical judgement or responsibility. Scientists do not always work as well as they could or should. Scientific theoreticians often make unwarranted pronouncements upon the nature of reality, which are regarded with suspicion especially by many of their experimentalist colleagues. And entire scientific communities may enter into uncritical groupthink preventing the emergence of more coherent alternatives (consider, for example, the 'central dogma' of molecular genetics that held back considerations of epigenetic inheritance, or the uncritical acceptance of Newtonian absolute space and time). Philosophers can and should ask critical questions concerning the reality of the items found in current scientific ontology.

Having articulated pragmatist notions of realness and realities, in the next chapter I will move on to discuss what it means to make true statements about realities. (If you are reading at the surface level you may now want to jump ahead to Chapter 4 from here.) In the rest of this chapter I want to consider more carefully several aspects of the practice of pragmatist metaphysics, outlining some key steps in the practice-based building of realities. Section 3.2 will discuss further how the mind-framing of entities is done in our activities and how the process of conceptual development unfolds. Section 3.3 will discuss the processes by which the mind-framed entities may be validated as realities. These discussions will give some detailed illustrations of how concepts, materials, experience and aims develop in full mutual entanglement. The kind of pragmatist metaphysics that I am proposing here will naturally lead to ontological pluralism, as shown in Section 3.4. Pluralism is a key aspect of the whole outlook on knowledge that I am advocating in this book. Finally, Section 3.5 will challenge a common reductionist ontology of physical composition that provides a strong source of resistance against pragmatist and pluralist metaphysics.

3.2 How Mind-Framing Works

So far I have rather abstractly advanced the notion that realities are mind-framed yet mind-uncontrolled entities. Now I want to spell out this idea in more concrete detail. The first step is to think more carefully about the actual processes by which the mind-framing of entities works. Realities are framed to fit what we do, and we need to understand how we actually create concepts and use them in our epistemic activities. The mind-framing of entities begins with ontological principles that are adopted because they are necessary for the performance of certain types of activity; some of the most fundamental features of the realities in our lives have their origins as conceptual prerequisites of our actions. Here I am building on Kant's fundamental insight that the mind provides certain a priori principles that frame experience. But the mind-framing of entities also has a more conscious and deliberate aspect of concept-design. We introduce new concepts and develop existing ones, in coordination with other existing and developing concepts, for various specific purposes. As suggested by C. I. Lewis, Michael Friedman, Sami Pihlström and others in a revision of the Kantian insight, we have a choice of which a priori principles to adopt. This element of freedom becomes obvious when we consider the processes of concept-development in science, as I will illustrate with the cases of the establishment and progressive development of the concepts of 'temperature' and 'acid'.

Mind-Framing and the A Priori

It is now time for me to say more precisely what mind-framing means. In relation to 'mind' I want to be quite liberal and consider, following Niiniluoto (2014, p. 160): 'perspective, point of view, practice, discourse, linguistic or conceptual framework, scientific paradigm, language-game, form of life, tradition, and style of thinking'. For my purposes, the important general point is that we use concepts in order to frame entities. But what, exactly, is 'framing'? To 'frame' means various things in ordinary English, and the meaning I intend is something like 'to formulate; form or articulate' (definition 2 in the Google Dictionary); this is also in line with the meaning of the noun 'frame' as 'a basic structure that underlies or supports a system, concept, or text' (definition 3).

But doesn't the image of framing actually support the correspondence-realist view of reality, if the entity already exists well-formed and the frame just encloses it, as with a picture-frame? That is not how I intend the term 'framing'. Going back to the etymology can help dislodge some unhelpful intuitions.[8] In Old English, *framian* meant 'to be useful', which then evolved into the Middle English meaning of 'to make ready for use' – a reassuring word-origin for pragmatists! An important case of 'making ready' was to 'prepare timber for use in building', which then led to the idea of the timber frame of a building, hence the basic structure of something. And then the thing one puts around a painting was *metaphorically* called a frame! It might be more useful, for our purposes here, to think in terms of the framework of a house. The frame does not contain or enclose the house; rather, it is an integral part of the house, without which the house would not stand.

Mind-framing is the specification of an entity in a form that can be handled by the mind. My view about mind-framing amounts to a denial of what Sam Page calls 'individuative mind-independence' (2006, p. 327): 'To say that the natural world is individuatively independent of us is to say that it is divided up into individual things and kinds of things that are circumscribed by boundaries that are totally independent of where we draw the lines.' In Page's

---

[8] The etymology is quite intricate. The account I give here is only one strand, extracted from the discussion in the Google English Dictionary online, provided by Oxford Languages.

terminology, I want to say that all entities are individuatively mind-*dependent*. This may sound like a wild metaphysical claim, but it really is just a productive tautology that is designed to shift our thinking in a certain direction: nothing should be called an 'entity' if it is not individuated.

But if realities are not controlled by the mind, how exactly is it that the mind 'frames' them? Kant gives us the most productive jumping-off point with his recognition of the a priori dimension of empirical knowledge, which in my view remains one of the greatest lessons in all of philosophy. And I think the full significance of Kant's insight can only be recognized when we consider how the a priori operates in the context of action, following the development of Kantianism in the directions proposed by Grene (1974) and Pihlström (2003; 2009). All human perception, thinking and communication take place within the confines of certain a priori concepts and principles, in terms of which we perceive, think, talk and act. These principles guide the mind-framing of reality at the most fundamental level. Where we have to depart from Kant is his insistence on the apodictic certainty of a priori judgement. Rather, what happens in mind-framing is a *suggestion*, a proposal to engage with prospective realities in a certain way. We start by postulating a certain type of reality, to see if such a conception can frame coherent experience. The a priori is what the mind imposes on experience, even though its validity is not guaranteed and only achieved through operationally coherent activities. But mind-framing is not a matter of random conjecture. Our starting-point is strongly constrained by the evolutionary path that humans have taken, in a direction that is positively adaptive on the whole. We have predispositions to think in certain ways, and those predispositions are also linked with our bodies, since we have embodied minds.

Recognizing the lack of absolute and eternal certainty in the a priori does not mean dismissing the very different roles that a priori and a posteriori judgements play in any given system of practice at a given stage of its development. This is one place where Quine's point about the in-principle holism of knowledge has often been taken in an unhelpful direction. What we need to recognize is that *in any given situation* some propositions are treated as empirical hypotheses open to testing, and others are taken for granted and protected from falsification. The latter are operative in the process of mind-framing; they are the principles that enable us to conceptualize and identify the entities we want to use and investigate. These principles are neither unalterably fixed, nor simply dispensable like ordinary empirical hypotheses. David Stump (2015) has given a very helpful characterization of such principles as 'constitutive' (instead of a priori), and made an informative survey of various philosophers who have recognized them.

Activity-Based Ontology

The first stage of mind-framing stems from the types of mental–physical activity that we are poised to take in the regular course of life. In previous work (Chang 2008; 2009a) I proposed that we should recognize a certain class of **ontological principles** (or, metaphysical principles) as necessary conditions for carrying out certain types of epistemic activity. These are the most fundamental principles with which we frame reality, and they are a priori commitments often made without explicit agreement or even articulation, simply through our deciding to undertake certain activities. If we did articulate these commitments, they would say: '*If* we want to engage in a certain type of activity, *then* we have to presume the truth of some particular metaphysical principles.' What we have here is a quasi-Kantian *conditional* or *contingent* transcendental argument – laying out the necessary preconditions for an activity that we engage in.

For example, if we are to engage in the business of making inductive predictions, we must take it for granted that the same conditions will results in the same outcome. Let us call this the 'principle of uniform consequence'.[9] An attempt to justify this principle by itself is futile, and that is why the 'problem of induction' is not solvable. It is correct to call induction a 'custom' as Hume did, but that misses the most important aspect of the situation. If we do participate in the form of life in which we predict what is going to happen next on the basis of what we have experienced before, then the principle of uniform consequence becomes an a priori principle. We can deny this principle, but then it would make no sense to attempt inductive prediction. What underlies the sense of necessity here is a pragmatic–hermeneutic kind of impossibility of doing without something.[10] The denial of an ontological principle while we are engaged in the activity that requires it would generate a sense of *unintelligibility*. What is involved here is just the kind of pragmatic sense-making, doing things designed to lead to the achievement of our aims, that I discussed in Section 1.4 in relation to the hermeneutic dimension of operational coherence.

---

[9] I would maintain that even in 'material' inductions as John Norton (2021) conceived them, a local version of the principle of uniform consequence is in action. And even if what we are making is probabilistic predictions, what we are doing is applying the principle of uniform consequence to groups of events. Determinism is a commitment to engage in prediction in every individual case.

[10] I want to argue that other types of impossibility are actually grounded in pragmatic impossibility, being metaphorical extensions of the latter. This is why it would be futile to try to analyse pragmatic impossibility further. In working out the notions of necessity and possibility sketched here, I wish to build connections to Roberto Torretti's ideas on the subject (1990, ch. 5).

Table 3.1. *A partial list of activity–principle pairs*

Activity-type	Ontological principle
Inductive prediction	Uniform consequence
(Contrastive) Explanation	Sufficient reason
Narration	Subsistence
(Linear) Ordering	Transitivity
Voluntary action	Agency
Intervention	Causality
Empathizing	Other minds
Individuation	Identity of indiscernibles
Testing-by-overdetermination	Single value
Assertion	Non-contradiction

For each well-defined type of activity, there is an associated ontological principle that makes it performable and intelligible. Table 3.1 gives a list of some important pairs of activity-type and ontological principle, and I will explain each item very briefly here. If we want to explain why something (as opposed to something else) happened, we have to assume that when there is an observed difference, there is a reason behind it; this may be considered a weak version of the principle of sufficient reason. The activity of narration requires that the stories we tell have subjects whose identities last through time, which 'house' the changes that are narrated. Paradoxically, without postulating something that lasts, it is impossible to describe any change. If we want to put a set of entities in an ordered sequence, we must assume that the relation that forms the basis of ordering is transitive. Engaging in voluntary action makes no sense unless we presume that our will directs certain parts of our bodies to move in certain ways. Intervening in the world with our own actions requires a presumption that our actions do make other things happen. In the activity of understanding other people's intentions and emotions, one needs to presume that they possess intentions and emotions like oneself. The identification of an object or a property as a distinct thing depends on an ontological principle close to Leibniz's principle of the identity of indiscernibles.[11] (The principle itself can be divorced from various uncertain uses that Leibniz and others have made of it.) When it comes to physical properties of objects, there is what I have called 'the principle of single value', which dictates that it can have no more than one definite value in a given

[11] I thank Roberto Torretti for this suggestion, as well as much inspiration and detailed discussions in the development of my thinking about ontological principles, even though he advised against using the term 'ontological' in this context.

situation (e.g., a stick cannot be 2 m *and* 3 m long). This principle is a prerequisite for the activity of 'testing-by-overdetermination', in which we determine the value of a quantity in two different ways (e.g., by prediction and observation); if the values match, that gives credence to the basis on which we made the two determinations.[12] Even the logical principle of non-contradiction may be an ontological principle, associated with the epistemic activity of asserting a proposition. Asserting something makes no sense unless we refrain from denying what we have just asserted. This one-to-one pairing of activity-type and ontological principle might seem too neat and contrived, but it makes sense considering that the ontological principle and the activity-type partially constitute each other. And the one-to-one correspondence only applies to the most basic types of activity; concrete and complex activities will require multiple principles.

It is satisfying to see how our basic metaphysical conceptions arise from the way we engage with the world. The most fundamental part of ontology is not abstracted from what we passively observe; rather, it emerges and becomes established as an essential ingredient of our coherent activities. Our inclinations to carry out certain types of activity begin to form the basic ontological shape of the world we live in. It is as Bergson said: 'The bodies we perceive are, so to speak, cut out of the stuff of nature by our *perception*, and the scissors follow, in some way, the marking of lines along which *action* might be taken' (Bergson [1907] 1911, p. 12; emphases original).

Freedom in Further Framing

The next stage of mind-framing comes when we consciously create or develop specific concepts in order to use them in various concrete activities. With the sort of activity-types I have been discussing so far, declining to engage in them would really make one depart from commonly recognized human forms of life. But there are also important conceptual–pragmatic choices to be made at less fundamental levels, and in such cases the freedom inherent in the process of mind-framing becomes much more visible. Especially in the long-term development of science, we witness a great deal of freedom being exercised in concept-creation and concept-development.

I believe that Kant made one major error, namely his commitment to universalism (cf. Niiniluoto 2014, p. 160). Regrettably he fell into the trap of regarding the trusted systems of knowledge of his age, including Euclidean

---

[12] For a more careful exposition of the principle of single value, see Chang (2004, pp. 90–1; 2008; 2009a).

geometry and Newtonian mechanics, as universally and necessarily valid. Looking at Kant from a little historical distance now makes one thing clear: his location in Newton-enraptured eighteenth-century Europe must have exerted a strong hold on his imagination. Take Philipp Frank's lament about the poverty of philosophical opposition to new scientific ideas: what parade as deep metaphysical truths are often simply 'petrified' remains of outdated scientific theories (Frank 1949, pp. 207–15). Metaphysical principles can and do change with the development of science. This is as acknowledged by a string of neo-Kantians ranging from William Whewell in the mid-nineteenth century to Michael Friedman in our time.

My chief inspiration here is Lewis, whose pragmatist notion of the a priori was explained systematically in his now-forgotten masterpiece of 1929, *Mind and the World-Order*. Lewis once reportedly declared: 'I am a Kantian who disagrees with every sentence of the *Critique of Pure Reason*' (quoted by Beck 1968, p. 273). The core of Lewis's disagreement with Kant was his denial of the existence of synthetic a priori judgements. Lewis stressed the great importance of a priori elements in knowledge, but argued that they were always analytic: *'The a priori is not a material truth, delimiting or delineating the content of experience as such, but is definitive or analytic in its nature'* (Lewis 1929, p. 231, emphasis original). For Lewis, all a priori principles follow from the nature of the concepts that we choose to craft and use:

> The paradigm of the *a priori* in general is the definition. It has always been clear that the simplest and most obvious case of truth which can be known in advance of experience is the explicative proposition and those consequences of definition which can be derived by purely logical analysis. These are necessarily true, true under all possible circumstances, because definition is legislative. (ibid., pp. 239–40)

Therefore, 'the necessity of the *a priori* is its character as legislative act. It represents a constraint imposed by the mind, not a constraint imposed upon mind by something else' (ibid., p. 197). These thoughts form the core of Lewis's 'conceptual(istic) pragmatism': there are a priori statements, which are true by definition, inherent in 'conceptual systems'; these systems are constructed by us, and adopted on 'instrumental or pragmatic' grounds (ibid., p. x). We choose the conceptual system freely, but once we have chosen a conceptual system, within the system the a priori elements are analytically true. I see Lewis's legacy in Anjan Chakravartty's (2017) recent work on scientific ontology, which combines realism concerning ontology with voluntarism in epistemology.

As a prime example illustrating his points, Lewis discussed Einstein's definition of 'distant simultaneity' in special relativity. Einstein defined the simultaneity of two events happening at a distance from each other on the basis of the principle of the constancy of the speed of light. So, if the two events happen at locations A and B, and the observer is at the mid-point M between A and B, then the two events are simultaneous if light signals released at the time and place of each event's occurrence reach M at the same time (*local* simultaneity being taken as unproblematically meaningful and decidable). But what is the status of the assumption that the speed of light is always the same in all directions? Einstein himself explains: 'That light requires the same time to traverse the path $A \rightarrow M$ as for the path $B \rightarrow M$ is in reality neither a *supposition nor a hypothesis* about the physical nature of light, but a *stipulation* which I can make of my own free-will in order to arrive at a definition of simultaneity.'[13] Such stipulations are necessary in the specification of the objects of inquiry, as Lewis explains here (1929, p. 256):

> we cannot even ask the questions which discovered law would answer until we have first by *a priori* stipulation formulated definitive criteria. Such concepts are not verbal definitions nor classifications merely; they are themselves laws which prescribe a certain behavior to whatever is thus named. Such definitive laws are *a priori*; only so can we enter upon the investigation by which further laws are sought.

So Einstein framed simultaneity in a fundamentally different way from Newton, and in a distinctive way that was almost completely unprecedented. It really does not make sense to claim that Einsteinian simultaneity, or Newtonian simultaneity, or any other variety, is inherent in nature. Rather, Einstein showed us how to frame simultaneity, and time itself, and space, too, together with light (and its constant speed) all in a tight package. Whether there were realities that could be so framed, which obeyed the a priori rules that Einstein laid down, was a contingent matter. The answer was to be found by seeing if the concepts involved supported operationally coherent activities. (Such pragmatic validation of mind-framing is the subject of Section 3.3.)

Mind-framing is not a once-and-for-all legislation of concepts; this becomes quite evident if we pay attention to the history of science. A striking aspect of scientific inquiry, as with any progressive enterprise, is that it continually introduces new concepts used for the mind-framing of entities, and updates and elaborates old concepts, too. The concepts change and evolve, and often there are divergent paths of evolution. This gives rise to competing systems of mind-framing (or what Goodman 1978 famously called 'worldmaking'), and scientists make choices between them. As Lewis points out, 'there

---

[13] Quoted in Lewis (1929, p. 256), from Einstein (1961, p. 23); emphases original in Einstein.

will be no assurance that what is *a priori* will remain fixed and absolute throughout the history of the [human] race or for the developing individual'. While Friedman stresses the hidden continuity even in revolutionary change, and detects continuity and progress at the level of constitutive principles (Friedman 2001, esp. p. 66), Lewis sees the epistemic agent as possessed of much greater freedom: 'If the *a priori* is something made by the mind, mind may also alter it'; 'the determination of the *a priori* is in some sense like free choice and deliberate action' (Lewis 1929, pp. 233–4).

The Evolution of Scientific Concepts: Two Cases

We can gain more insights about the mind-framing of entities by examining some further concrete cases from the history of science in detail. There are numerous examples to choose from, but I will consider two that I have treated in some detail in previous works (see Chang 2016a for further examples). The first example is temperature (Chang 2004). Temperature is more of a property than an object, but when I speak of the mind-framing of entities I am taking 'entity' in a broad sense.[14] Let's start with the basic ontological principles relating to temperature. In building a quantitative concept of temperature out of the vague notion of hot and cold, scientists first of all presumed that temperature was a real physical quantity, subject to the principle of single value. The activity of testing-by-overdetermination (associated with the single-value principle) manifested itself particularly as the practice of checking thermometers for comparability: does a given thermometer always give the same temperature reading when placed in the same situation, and do different thermometers give the same value in the same situation? Similarly, the principle of transitivity was also applied to the temperature concept, assuming that objects could be put into a linear ordering by their temperature. This placed temperature onto an ordinal scale of measurement. Such stories are very common in the history of measurement: we propose the existence of a quantified property in nature, and try to find ways of getting at that quantity. In the conception of a measurable quantity, it is difficult to imagine the absence of mind-framing in terms of the single-value and transitivity principles.

Going beyond the conception of temperature as a measurable quantity in itself, an important step was to enrich and refine it in connection with

---

[14] I will use the term 'entity' to designate any kind of thing we may conceive and discuss that may be designated by a noun term. An entity is not necessarily a material thing; it can be an event or a process. It may even be abstract, or social or institutional. (I could also take 'object' in a similar way, but there are too many conflicting intuitions about 'object', so I try to avoid using that term altogether.)

other concepts. A major step, taken in the late eighteenth century, was to distinguish temperature from heat, and to clarify the relation between the two. The concept of heat had separately been developed as a quantity on a ratio scale (with a physically meaningful zero, and addition and multiplication operations). For a long time heat was also associated with the notion of the material substance called caloric, but this association fell away gradually. Temperature came to be linked to heat chiefly by means of specific heat, defined as the amount of heat required to raise the temperature of a substance by a unit amount. This link helped in establishing temperature as a quantity measurable on an interval scale, on which differences between two values are physically meaningful: it takes the same amount of heat to raise a given body by the same temperature interval (e.g., 10–20 degrees, or 40–50 degrees), if the specific heat of the body is constant. The next major step was to link temperature and heat with mechanical concepts such as velocity, kinetic energy and mechanical work, through the new theories of thermodynamics and statistical mechanics. Temperature acquired new meanings, on the one hand as 'absolute temperature' in thermodynamics, later linked with entropy as well (see Chang 2004, ch. 4; Chang and Yi 2005), and on the other hand as something proportional to the average kinetic energy of molecules. Each of these developments settled down in the form of a definitional proposition taken as a priori in its own context, and thereby making a deep change in the framing of 'temperature'.

The other example I want to discuss is the concept of 'acid' in chemistry (see Chang 2016a and references therein). As a substance term, 'acid' specifies a very different kind of entity from property-terms like temperature. Here we start with framing in terms of the ontological principles that typically pertain to substances, such as subsistence, causality and the identity of indiscernibles. In the earliest part of the history, the significant move that chemists made was to take 'acid' as a subsisting thing in its own right, rather than treating acidity as a transient property exhibited by various substances. But what kind of thing is it? How do we make the term 'acid' refer to some recognizable set of stuff? The obvious step for the working chemists was to choose a set of typical properties and behaviours as defining characteristics of an acid: sour taste, corrosiveness (especially in relation to most metals), and the ability to alter the colours of various indicators (litmus, juice of violets, turmeric, etc.) in specific ways. Almost as important was the fact that acids and alkalis neutralized each other so that they lost their typical properties. These steps constituted the inevitable intensional route to reference-fixing, which I discussed in Section 2.3.

After this property-cluster definition of acid settled down well enough, many attempts were made to identify the 'essence' of acidity, on the assumption that there was one fundamental characteristic of acids that was responsible for all their other properties. This was an important additional layer of mind-framing that acids went through: it was not necessary to conceive of acidity in such an essentialist way, but that is the choice that chemists made – not to ask *whether* there was an essence to acidity, but to ask *what* that essence was. Perhaps the best-known attempt was by Lavoisier, who thought that oxygen was the 'principle' of acidity (and coined the term 'oxygen' from Greek roots to mean 'acid-maker'), but there were various other candidates proposed as well. All these early attempts failed, and success only came in modern times, when first Svante Arrhenius, and then J. N. Brønsted and T. M. Lowry developed the notion that it was the ability to give up a hydrogen ion that defined an acid. And it was only with this advanced theoretical concept that a convincing quantification of acidity began, via the pH concept and the glass electrode enabling the construction of pH meters (Ruthenberg and Chang 2020). But almost simultaneously the work of Gilbert Newton Lewis introduced the notion of acid as the acceptor of an electron-pair, which could be taken as a broader theoretical category encompassing the Brønsted–Lowry acid concept, but not having a clear link with the pH measure.

Note the variable and unsettled course of development in the framing of 'acid', already evident in the very brief sketch I have given here. This is a good illustration of the shakiness of the correspondence-realist notion that there is some well-defined thing out there which our concept can simply point to or 'latch on to'. Through the changes mentioned above, not only the intension but the extension of the concept 'acid' changed significantly. I would say that each notion of acid mentioned above did frame *some* real entity; however, the various entities framed by the various concepts are not at all one and the same thing. And it is difficult to say which concept is the best one. Perhaps some of the earlier ideas can be discarded safely enough in the context of modern chemistry, but it is very difficult to choose between the Brønsted–Lowry concept and the Lewis concept. The Lewis concept is the most sophisticated one theoretically, but it is basically not measurable. Most of the experimental activities concerning acidity are carried out on the basis of the Brønsted–Lowry concept, and there is nothing theoretically *deficient* about that concept. In practice, modern chemistry retains both of these concepts. There is no such thing as mind-unframed 'acid', and we can and must choose which mind-framed 'acid' we want to engage with.

3.3   The Achievement of Reality

We are free to make up concepts as we like, but they need to be validated pragmatically by supporting coherent activities. Only then are we in possession of mind-framed *realities*. Achieving reality requires a dynamic working-together of experience, action and concepts in close interconnection and coordination. I continue with the cases of temperature and acid to illustrate the process of validation. The character of reality achieved in this way is quite strikingly different from the usual notions of reality. In my conception, reality (real-ness) is a matter of degrees, and it is also domain-specific. When we create a new concept that turns out to be able to support coherent activities, we have a new reality. And when we make further coherent activities relying on an existing reality, we make it more real than before. Reality is an achievement made through well-designed concepts and activities. But as with any other achievements in life, whether our attempt to achieve reality succeeds or not is ultimately not in our control.

The Pragmatic Validation of Concepts

Having considered how the mind frames realities, let us turn to the aspects of realities that the mind has no control over. Experience is the ultimate source and touchstone of our knowledge about mind-uncontrolled realities. We could even say that 'experience' is the generic name that we give to our encounters with mind-uncontrolled realities. It is through experience that we have 'contact' (metaphorically and literally) with realities. These are tautologies that express the empiricist outlook. Even though the objects of experience clearly embody a priori elements, they are realities as long as they support operationally coherent activities. So I return to my philosophical idiom that realities are mind-framed but not mind-controlled.

The question now is how our freely chosen framing of realities can be validated by experience, so that our proposed entities may be shown to be realities (or not). Freedom and choice do not mean arbitrariness. A quick example will illustrate the point. In his attempt to reform the foundations of classical mechanics, Ernst Mach ([1889] 2013) eliminated the traditional Newtonian concept of force, which he considered too metaphysical. But this also removed Newton's second law *(F = ma)* and, along with it, an obvious way to define mass, as *F/a*: the strength of force applied to an object divided by the acceleration resulting in its motion. To fill this gap, Mach offered a new definition of mass: if two objects are allowed to interact with each other and they undergo accelerations as a result, the ratio of their masses is the inverse of the ratio of the (magnitudes of) accelerations induced. In a formula: $m_1/m_2 = a_2/a_1$. So far, it is just a matter of laying down a definition. But suppose we have three objects in the system. Then we would have $m_1/m_3 = a_3/a_1$ and $m_2/m_3 = a_3/a_2$. Now we have a constraint on the acceleration values, if mass is to obey the principle of single value (see Section 3.2). To take the simplest possible case, suppose that objects 1 and 2 induce the same magnitudes of acceleration in each other, and objects 1 and 3 do so, too. Then, by Mach's definition, objects 1 and 2 have the same mass, as do objects 1 and 3. Then objects 2 and 3 must have the same mass, too, which implies that they must induce the same acceleration in each other. But will that be borne out by experiment? Mach himself notes: 'No *logical* necessity exists whatsoever, that two masses that are equal to a third mass should also be equal to each other' (ibid., p. 219; emphasis original). What should we do if a_3 and a_2 come out different when we do the experiment? We will not be able to do much operationally coherent physics in such a situation, and we would be forced to reject the Machian definition as unworkable. So we see that Mach's definition actually contains a hypothesis about how accelerations will go in physical situations. Definitions and Lewisian a priori principles are not mere tautologies; whether they are apt or not depends on how things turn out empirically.

Coherence-Building

The pragmatic validation of concepts, when it works out, is an iterative process of building operational coherence. In order to illustrate this point, I will continue with the two examples introduced in the previous section. In my discussion of the temperature concept in the last section, I mentioned that the presumption of single-valuedness demanded comparability in thermometers. To us moderns it is difficult to imagine what a great challenge it was to ensure comparability in thermometers, but the fact is that for nearly two and a half centuries thermometers were made without convincing comparability, until the monumental and painstaking work of Victor Regnault in the middle of the nineteenth century. What Regnault's work revealed was that only thermometers filled with air (or one of a few other gases) exhibited a sufficient degree of comparability. The much-loved mercury-in-glass thermometers failed the test of comparability, and thermometers filled with any other liquids (including alcohol) were even worse. In other words, temperature-measurement was a coherent activity only when it used a few particular types of thermometric fluid.

There was also a challenge in identifying the thermometer-based concept of temperature as a quantified version of people's sensations of hot and cold. The indications of thermometers often went against qualitative perceptions, and such disagreements had to be accommodated by finding plausible explanations for them. For example, it feels so cold even though the temperature isn't so low, because of wind chill. I am having chills even though the ambient temperature is high according to the thermometer, so I must be getting a fever (a suspicion that I confirm with the thermometer). And so on. With such challenges overcome, the thermometer-based temperature concept was able to enhance the operational coherence of a whole host of activities in many areas of life, including horticulture, brewing, clinical medicine and experimental chemistry and physics, thanks to the greater precision and reliability in the judgement of temperature enabled by the thermometer. Thereby temperature as a quantity became a firm reality.

A different kind of developmental pattern and challenge can be seen in the development of the temperature concept in relation to the concept of heat. Specific heat initially arose as a vague material notion of 'heat capacity' (or capacity for holding caloric), with the density of caloric representing temperature. The caloric-based thinking did not survive in the end, but the more phenomenological concept of specific heat as the ratio between heat input and temperature increase survived robustly, and supported a set of very coherent calorimetric activities involving the operations of mixing various substances at different initial temperatures and predicting and measuring the temperature of the resulting mixtures. But a major apparent incoherence loomed: in many situations the addition or subtraction of heat from a body did not change the temperature at all – is the specific heat infinite in such cases? Such a consequence was avoided by the invention of the concept of *latent* heat: let's postulate that any heat input that does not serve to raise the temperature of the receiving body goes into a latent (non-sensible) form; this made a lot of sense, since latent heat could also be understood as the cause of observable changes in the state of the receiving body – most notably melting and boiling. And the reverse changes of state (freezing and condensation) were duly seen to release the latent heat back into sensible form, again without a change in temperature. So, latent heat was seen to be real, while the coherence of a whole range of activities in thermal physics was maintained and enhanced.

With such an impressive and wide range of coherent activities relying on it, the reality of temperature as a quantified property became very clear. By now its reality is exhibited in almost every area of science, industry and medicine, and it seems that every process in nature is affected by temperature.

It should be stressed again that the concept of temperature has continually grown and changed over the centuries, within a complex and growing network of concepts. This also means that the reality designated by the temperature concept is complex and changeable. An extreme instance of this is the fact that the modern thermodynamic concept of temperature allows the reality of *negative absolute* temperature, if there were a physical system whose entropy decreases (becoming more orderly) when it absorbs heat. But negative absolute temperature is clearly not a possibility under the kinetic theory of heat, according to which absolute temperature is proportional to the average kinetic energy of molecules, which cannot take on a negative value.

If we look at the case of acid, some different issues emerge. Concerning material substances there is a patchier track-record in scientists' postulations settling down as realities, compared to measurable quantities. Many presumed entities that enjoyed scientific popularity for a time have come to be considered unreal: the four Galenic humours, caloric and phlogiston, the aether, and many other items that feature in Laudan's list supporting the pessimistic induction. In the case of acid, too, it might have been questioned along various points in its history whether there was really such a thing. In the property-cluster stage of the concept, it was not clear whether all the properties were exhibited by all the acids. To put it in Richard Boyd's terminology (1999): is the property-cluster in question homeostatic? Do all the key properties of an acid always go together? If not, a whole range of activities from classification to prediction based on the concept would lack coherence. For example, 'carbonic acid' (CO_2 in modern terms, or rather, its combination with water, H_2CO_3) has no sour taste but will turn litmus red. Given such gaps, the reality of 'acid' was not convincing.

In the stage of development in which people looked for the essence of acids, different kinds of activity came into focus. If there really is a 'principle' of acidity (such as oxygen), then it should have been possible to apply it to other substances to turn them into acids. Sometimes this did work out, as many products of combustion turned out to be acidic, at least when dissolved in water; for example, carbon dioxide (CO_2) produced in the combustion of organic substances formed carbonic acid when it met water. But not all products of combination were acidic. Another thing that ought to work if acids contain oxygen is to extract oxygen from known acids, but sometimes this also turned out to be impossible, as in the famous case of 'muriatic acid' (hydrochloric acid, HCl). So the activities of acid-production by oxygen and oxygen-extraction from acids were seen to lack full coherence. In the end chemists abandoned not only Lavoisier's oxygen theory of acids, but also the general notion that there was an essential substance that conferred acidity on other substances. This is a good illustration of how our concepts may not turn out to designate realities. In contrast, the rendition of acidity as a disposition worked out much better. According to the Brønsted–Lowry conception, an acid is a substance capable of donating a hydrogen ion, and on that basis a whole range of coherent activities can be performed, ranging from the definition and measurement of pH to the understanding of neutralization reactions as the meeting of hydrogen and hydroxide ions (H^+ and OH^-) to form water. So we say confidently that Brønsted–Lowry acids are real.

The Character of Reality

Having tried to convey a concrete sense of how entities are framed and their reality (real-ness) is achieved, let me now consider more carefully the general character of reality achieved through the establishment of coherent activities. The first thing to note is that reality, or even our knowledge of it, is not simply related to observability or any other kind of possibility of immediate access. There is no fundamental difference between the processes we use to determine the reality of observable and unobservable entities. To return briefly to the case of temperature: scientists were able to establish quite convincingly the reality of something as theoretical and removed from direct observation as Kelvin's absolute temperature, which is defined in terms of thermodynamic theory in the setting of an ideal Carnot engine, which no one has ever been able to make to any tolerable approximation. The same establishment of reality can work out for more esoteric, theoretical and unobservable entities like quarks and dark energy, too.

The next point to note is that being real is a matter of degrees, as I have tried to make clear in the examples discussed above. At first glance this will seem absurd, but I hope that the sense of absurdity will dissipate on careful consideration. We might start by noting that in colloquial speech we do easily attach degrees to real-ness: 'the terrorist threat in London is still *very real*, even though less media attention has been given to it lately'. An entity should be considered real to a higher degree if it supports a larger number of activities that are operationally coherent. In addition, operational coherence itself is a matter of degrees, so the real-ness of an entity is higher if each activity it supports has a higher degree of operational coherence. And it is real-ness itself that is a matter of degrees, not just our knowledge of it. Still, 'degrees of reality' will sound very strange to the ears of those schooled in contemporary metaphysics (though it was a notion sometimes entertained traditionally): isn't real-ness the same thing as existence, and doesn't existence have to be an all-or-nothing affair? Can something be just a little bit real, or partly exist? I think we would do well to find ways of accepting such a notion.

Existence is seen as a black-and-white issue only because we are accustomed to thinking about extremely clear-cut cases with clear meanings. There is no reason to maintain that an ill-defined entity either exists, or not. Consider: 'Is there any moral turpitude in this man?' Even in science, existence may not be a black-and-white matter. Do light rays exist? They do, in the sense that we can track straight, refracted or reflected paths of light. But they also do not exist, in the sense that there is no material body in the shape of lines that can be exhibited along the path of light. Familiar and concrete cases like the Loch Ness Monster may not be a simple yes/no matter, either. What if there is a creature in the lake that is a lot like what people have described, but not quite? Or what if it is a visual effect that surely looks like what people have reported, but with no material substance behind it? We should say that there is *some* reality to the monster, since some activities involving the monster will be quite coherent if there is a Nessie-like creature or a good visual effect mimicking it.

The reality of entities is not only a matter of degrees, but also something pertaining to specific domains. 'Domain' here may be a spatio-temporal region, but more generally I intend the term to refer to all kinds of conditions that affect the coherence of an activity, pointing to a rather general type of context-dependence. Real entities are only real in their proper domains, not everywhere. Newtonian point-particles are real enough in situations where Newtonian activities are coherent (including 'rocket science', solar-system celestial mechanics and pendulum motion, just to take a few examples). Quantum wavefunctions defined by the Schrödinger equation are real when it comes to electrons in atoms, and not so much when it comes to protons and neutrons in the nuclei. Old-fashioned light rays are very real where geometric optics works, but clearly not when we are doing the double-slit experiment. Constellations are real within traditional positional astronomy, but not in modern cosmology. Atomic weight as a fixed number unique to each chemical element is real in the construction of the periodic table, but not in nuclear physics.

But you might object: how can a given entity be real in one domain and not in another? Suppose we refract a ray of light with a prism to direct it on to a metallic surface, from which it will cause an electron to be ejected by the photoelectric effect. It seems that my notion of reality will force me into the absurd view that what goes through the prism is a light ray, which somehow turns into a photon-bundle as it hits the metal surface. Shouldn't we rather learn to make a unified account in terms of what is actually real everywhere (namely photons, not light rays)? But such a view is based on complacency: you think you have the true picture of Reality, which contains photons, not light rays. But when you say 'photon', what exactly do you have in mind? You probably think it is a parcel of energy in the amount of hv, where v is the frequency of light and h is Planck's constant. But shouldn't you be thinking of what quantum field theory or even superstring theory says light is, or rather, whatever the finished-and-accepted 'theory of everything' will say light is? So, in principle, we can't pronounce at all about photons until physics is all finished. But let's look at the reality of our practices. Whatever our 'final theory' may say, we do already have many coherent practices involving light rays, photons and also electromagnetic waves. But can these different entities all be real? I will address this question of ontological pluralism further in the next section.

3.4   Ontological Pluralism

The message from the discussions given so far in this chapter is clear: if we want to think seriously about the nature of realities, we should take heed of what happens when scientists and other investigators try to craft concepts that facilitate coherent activities. When we do, one significant thing we learn is the pervasiveness of plurality: diverse types of ontology support various sets of coherent activities, even within the realm of science, and even within specific areas of science. This pluralist lesson goes against a deeply ingrained metaphysical picture, in which the universe has one correct inventory of things. There is no convincing justification for this monist ontology, either from the track-record of science or from general philosophical considerations. Pluralist ontology becomes easily acceptable when we move away from the fallacy of pre-figuration, the notion that well-formed real entities simply exist 'in the world' (see Chapter 2). Real entities are mind-framed, and there is no absurdity in allowing that many different sets of real entities are operative in a given field of science. I advocate ontological pluralism (to be discussed further in Section 5.4): it is beneficial to encourage multiple ontologies, each of which can facilitate coherent epistemic activities.

Ontological Pluralism and Its Traditional Sources

If we accept that operational coherence provides a good criterion of reality, then it will be difficult to avoid accepting a diverse array of ontologies. If we add the coherence theory of reality to epistemic pluralism, we are bound to get a modest, practical sort of ontological pluralism. In my first statement of pluralism I explicitly limited myself to epistemic pluralism, on the ground that experience did not teach us sufficiently well about metaphysics (Chang 2012a, ch. 5). However, over the years I have gradually come to see the force of Cartwright's view that the successes of certain practices do give us credible indications about ontology. Here I accommodate her insight in my own way, by recognizing that what we *mean* by realities is the entities that facilitate operationally coherent activities. From the diversity of successful local practices, Cartwright (1999) takes a single picture of the world that is variegated and inhomogeneous, and of Nature that works like our own artful practices of modelling (Cartwright 2019). What I take is a plurality of ontologies (each of which may be 'dappled', or not), in a way similar to Annemarie Mol's (2002) identification of the overlapping and interacting multiple ontologies emerging in medical practice, and Chakravartty's (2017, p. 190) 'pluralism about packaging' and 'pluralism about behavior'.[15]

My conception of reality is stringent, but it is also permissive – or 'promiscuous', in Dupré's provocative phrasing. The criterion of operational coherence does rule out many things as candidates for reality, but also rules in a whole variety of other things. In the absence of what else we might operationally mean by 'real', and with the recognition that a concept of reality is not something we should try to do without, I propose that we muster the metaphysical courage to admit that there may be different systems of real entities, even within what we might be compelled to regard as one and the same domain. My thinking is in the tradition of naturalism, treating ontology as a field of study best approached through empirical inquiry. Now, if any entity that plays an indispensable role in coherent activities is taken to be real, then it opens up the possibility that all sorts of entities may be real, all at the same time. This is a contingent matter, but what we have seen in the history of scientific endeavours is that very different ontologies have indeed supported various successful scientific practices. This issue becomes especially acute if we dispense with the presumption of ontological reductionism, as I will propose in Section 3.5.

In embracing and developing ontological pluralism, I want to start by noting that it is a position that has been advocated by some eminent and sober modern philosophers (see Stump 2020 for a helpful survey). It is very well known that Rudolf Carnap argued that we have a choice of conceptual frameworks, each with its own fundamental ontology:

---

[15] I thank Helene Scott-Fordsmand and Brooke Holmes for introducing me to Mol's work.

Let us grant to those who work in any special field of investigation the freedom to use any form of expression which seems useful to them; the work in the field will sooner or later lead to the elimination of those forms which have no useful function. Let us be cautious in making assertions and critical in examining them, but tolerant in permitting linguistic forms. (Carnap 1950, p. 40, emphasis original)

Quine's doctrine of 'ontological relativity', briefly mentioned already in Section 2.3, is bound to have pluralist implications. Putnam's 'permutation argument', fully anticipated by Quine, can also easily be read in a pluralist vein, and Putnam was explicit in his lasting commitment to ontological plurality even after he abandoned 'internal realism'. In response to Maudlin (2015), whose argument for metaphysical realism was discussed in Section 2.2, Putnam (2015b, p. 503) admits that he is a metaphysical realist, but proposes a 'sophisticated metaphysical realism' based on an acceptance of ontological plurality: 'the same state of affairs can sometimes admit of descriptions that have, taken at face value, incompatible "ontologies," in the familiar Quinean sense' (ibid., p. 506).

Among contemporary authors Niiniluoto (2014, p. 160) is notable for proposing a 'principle of conceptual pluralism', which admits that 'all inquiry is relative to some conceptual framework'. With this principle, and an admission of Peircian fallibilism, he arrives at his 'critical realism', which modifies metaphysical realism so as to make it more operational in practice.[16] Kitcher (2012, pp. xxii–xxiii) proposes a modest sort of ontological pluralism as the first key features of 'the pragmatist reform of epistemology and metaphysics': 'whatever is independent of us might be conceptualized in many different ways'. That sounds innocuous enough for the metaphysical realists, but we should read on: 'In one sense there is only one world – the yet-to-be-differentiated source of our experience; in another, there are many worlds – the diverse articulated totalities of objects assorted into kinds that reflect all the ways in which divisions might be drawn.'

Equal-Opportunity Experimental Realism

So there have been plenty of in-principle discussions about alternate ontologies. I would like to focus more on how multiple ontologies *arise in practice*. In looking to practices for ontology I have already drawn inspiration from Hacking's experimental realism, and here I will do so again. There is one obvious objection to Hacking's position, which will turn out to be a blessing

---

[16] Niiniluoto's view is that Kant was driven to scepticism about knowledge of things-in-themselves because he didn't recognize pluralism, implying that pluralism allows knowledge of things-in-themselves. That is where I part company with Niiniluoto.

in disguise for pluralism. The objection is this: might we not *misunderstand* our experiments, mistakenly presuming that some non-existent entity is involved in them? This actually seems to happen with some regularity. The history of science is full of very successful practical interventions by experimenters who thought they were using entities that we now regard as unreal. The pessimistic induction from the history of science would seem to be just as deadly to Hacking's experimental realism as it is to standard scientific realism.

Let me illustrate the problem with some concrete examples. William Herschel discovered infrared radiation in 1800 by realizing that a thermometer inserted into the dark space beyond the red end of the solar spectrum detected a good deal of heating effect (see Hentschel 2002, pp. 61–4; Chang and Leonelli 2005). Herschel thought that he had successfully used a prism to separate out the rays of light and the rays of caloric, both contained in the sunbeam, directing the caloric rays onto the thermometer. If this doesn't qualify as Hacking-style 'spraying', I don't know what does. Now, doesn't such a case amount to a refutation of Hacking's experimental realism, and also my coherence theory of reality, since we now know that caloric isn't real? Meanwhile, among Herschel's contemporaries who opposed the caloric theory we find Count Rumford, often celebrated as the neglected pioneer of the kinetic theory of heat. Rumford held that heat consisted in the vibration of molecules, and even showed by experiment how much heat could be generated by friction. But most who praise Rumford's prescience do not realize that he also postulated the existence of 'frigorific' radiation, namely low-frequency waves emitted by cold objects, which have the effect of cooling down warmer objects at a distance. Rumford devised successful experiments to reflect and focus frigorific rays using metallic mirrors and cones (see Chang 2002).

Similarly, consider the infamous case of phlogiston (Chang 2012a, ch. 1, esp. pp. 53–4). The phlogiston theory made sense of a key aspect of age-old smelting techniques, in which calx (metal oxide, in modern terms) was transformed into metal by taking up phlogiston from a combustible (i.e., phlogiston-rich) substance such as charcoal. Even Kant greatly admired Georg Ernst Stahl's laboratory operations transforming one substance to another, and back to the original one, by giving phlogiston to it and taking it back out (Kant [1787] 1998, 108–9). Similarly, Joseph Priestley claimed to be able to manipulate phlogiston successfully. Matthew Boulton (James Watt's business partner) wrote excitedly to Josiah Wedgwood (the famous porcelain-maker) in 1782:

> We have long talked of phlogiston without knowing what we talked about, but now Dr Priestley hath brought ye matter to light. We can pour that Element out

of one Vessell into another, can tell how much of it by accurate measure is necessary to reduce a Calx to a Metal ... (Boulton quoted in Musgrave 1976, p. 200)

In the 1770s Priestley made oxygen through his attempt to 'de-phlogisticate' air by reducing a calx (rust) back into metallic form in an enclosed space; the air in that space would give up its phlogiston to the calx, restoring its metallic nature. This seemed to work out, and he obtained 'de-phlogisticated air'. This new gas supported combustion exceptionally well because it was eager to regain phlogiston, plentiful in combustible substances. Even less ambiguously, Priestley predicted that a calx could be reduced by heating in inflammable air (later called hydrogen), which he conceived as pure phlogiston. This experiment succeeded brilliantly. Did Priestley's successes, including even novel predictions, mean that he and his contemporaries should have granted reality to phlogiston?

Such cases are not just relics from bygone eras. Take, for example, orbitals in modern chemistry. Not only a great deal of theoretical explanation but numerous experimental interventions in modern chemistry rely on the general concept of orbitals, and also on a detailed knowledge of the number and shapes of various types of atomic and molecular orbitals. In addition, the idea of sequential filling of atomic orbitals explains very nicely, to a limited yet very significant extent, why the periodic table has the shape it has (see Scerri 2007). But orbitals inhabited by individual electrons have no reality if we take quantum mechanics literally, since all electrons are identical and they cannot be said to occupy different orbitals within a given atom or molecule (see Ogilvie 1990). Yet, I think it makes sense to attribute reality to orbitals on the basis of the successful chemical practices employing them, and that is certainly how very many chemists think.

All of these examples may appear to show that coherent experimental activities provide no guarantee of the reality of the entities that the experimenters themselves *presume* to be manipulating. You may be able to spray something without knowing much at all about what it is that you're spraying. So, are we back to square one, with Hacking's attempt to save realism all in vain? I think the way forward is to admit that caloric, phlogiston and such entities *are* real in their proper domains. That is much more defensible than the selective fault-finding mission, targeting activities involving entities that we currently think are non-existent. Not only is this strategy unprincipled, but it is a hostage to fortune as the scientific consensus shifts around, concerning which cutting-edge theory is the correct one. This is another instance in which the faith in science that I have critiqued in Section 2.4 can lead to unwarranted complacency. Without presuming that our current most popular theories tell us the truth about ultimate reality, how do we *know* that caloric and phlogiston and such things are *not* real?[17]

What I am proposing is a kind of liberal equal-opportunity realism. Let us grant reality to all entities that support coherent activities, where they do and to the extent that they do. It is useful to recall another part of Hacking's argument here. Facing those who would doubt that success in practical intervention can form a secure enough basis for our knowledge of unobservable reality, Hacking hits the ball right back to their court, by asking: why do you think *anything* is real? In an argument already alluded to in Section 1.6, Hacking points out that even the 'medium-sized dry goods' are only considered real because of our ability to handle them. This includes the visual–muscular coordination that is such an essential part of our normal acts of seeing.[18] So why not admit that phlogiston within its domain of coherent use is (nearly) as real as tables-and-chairs are in our daily lives? And here is a thought for the standard scientific realists who put their trust in the argument from success: we should be open-minded and generous to all investigators, by granting reality, provisionally and defeasibly, to the referents of whichever theoretical conceptions seem to lead to success. This is what we ought to do if we really take success as our only reliable guide in deciding what to be realist about.

Living with Plurality

Still, I don't imagine that standard scientific realists will immediately be happy with the liberal line that I have just taken. They will point out, quite rightly, that taking both caloric and phlogiston as real implies accepting two very different stories in the same domain, for example the science of combustion. So, will ontological pluralism end in contractions? And would that not destroy the overall cogency and authority of scientific knowledge? I will argue that this fear is exaggerated and that the pluralistic situation actually offers tangible benefits. Ontological plurality is a matter of contingent fact about our epistemic life: it just turns out that we can often engage in coherent activities by employing

---

[17] The same problem emerges if we try to take the pessimistic induction literally as an induction. If we *know* that the terms in Laudan's infamous list do *not* refer, that means we do know something about the shape of reality, and then the realists have won. The pessimistic 'induction' is only valid if it is taken as a *reductio*, as Psillos (1999, p. 102) points out.

[18] The classic experiments by Richard Held (1965) seemed to show that normal vision failed to develop when kittens were deprived of muscular activity moving themselves around. See Bermejo, Hüg and Di Paolo (2020) for a retrospective.

many different kinds of presumed entities in various situations. Even within a given domain, it often happens that there are multiple coherence-conducive ontologies that cannot easily be conceived in terms of each other – such as wave and particle, phlogiston and oxygen, caloric fluid and molecular kinetic energy, or electrons neatly pigeonholed into orbitals and a 'gas' of mutually indistinguishable electrons. If we accept my notion of reality, we should be open to the reality of sets of entities that are apparently mutually incompatible. In fact that would also be the case even for standard scientific realists who stay true to the spirit of the realist argument from the success of science, when equal success is achieved on the basis of each of the competing ontologies.

Above I mentioned some cases of plurality drawn from the history of the physical sciences in recent centuries. I predict that a more thorough and extensive look across various sciences and many other areas of life will reveal the co-existence of very different ontologies in *most* domains of activity and thought. For example, our legal thinking is mostly done in terms of individual persons and their actions, yet at the same time corporations are treated legally as persons. Most of us think about human actions in terms of free will and moral responsibility, while at the same time agreeing that the mind is only a manifestation of the molecular and electrical activities in the brain. As Arthur Eddington (1928, pp. 1–5) famously put it, 'there are duplicates of every object'. Sitting down to write his words, he was at his 'two tables': first, the familiar substantial thing, and then the 'scientific table', which is 'mostly emptiness' with 'numerous electric charges rushing about with great speed'. Would you, on reflection, deny reality to either of Eddington's two tables? Coming back to physics, currently we have the ontologies of curved space-time and quantum fields, sitting superimposed on each other as it were, with dark matter and dark energy tucked away somewhere in the picture.

It is instructive in this connection to hear C. I. Lewis again, this time on the progress of knowledge and conceptual change:

> New ranges of experience such as those due to the invention of the telescope and microscope have actually led to alteration of our categories in historic time. The same thing may happen through more penetrating or adequate analysis of old types of experience – witness Virchow's redefinition of disease. What was previously regarded as real – e.g., disease entities – may come to be looked upon as unreal, and what was previously taken to be unreal – e.g., curved space – may be admitted to reality. But when this happens *the truth remains unaltered and new truth and old truth do not contradict.* Categories and concepts do not literally change; they are simply given up and replaced by new ones. (Lewis 1929, p. 268; emphasis original)

Lewis does not draw an explicitly pluralist conclusion in this passage, but he takes two crucial steps towards it. First, he says that an old truth 'remains unaltered', while he expects conceptual change to continue, and along with it ontological change. Lewis considers it natural and right that we attribute reality to the entities that play significant roles in the conceptual schemes through which we live and learn at each stage of development. There is no final point or destination of development, which is to say that nothing we regard as real should be regarded as absolutely and exclusively and eternally real. That is also to say, to the extent that something is real now, it should not cease to be real just because it becomes necessary for us to deal with new experiences, for which we need a different conceptual scheme.

Second, by seeing that 'new truth' and 'old truth' do not contradict each other, he allows some logical breathing space for pluralism. It is not necessary, and in fact not often, that different ontologies directly contradict each other. Take the case of oxygen and phlogiston again. It is often thought that Lavoisier's oxygen-based chemistry proved that phlogiston did not exist, but that is too hasty. No plausible definition of 'phlogiston' and 'oxygen' can give us a logical deduction that 'phlogiston is real' implies 'oxygen is not real', or vice versa. So there is no direct logical contradiction in affirming that both entities are real, and in fact there were chemists who, in the midst of the Chemical Revolution, made coherent hybrid systems of chemistry which affirmed the reality of both, using oxygen in the tracking of weights and phlogiston in the explanation of (what we now recognize as) energy relations. (see Chang 2012a, p. 32, and references therein). The phlogistonist and oxygenist systems of chemistry did contain some mutually contradictory statements (such as 'Water is an element' in one and 'Water is a compound' in the other). However, if we sufficiently dissect the meanings of 'element' and 'compound' in those sentences, we find that semantic incommensurability prevents any direct contradiction (see Chang 2012a, pp. 208–12; cf. Goodman 1978, p. 110). I will comment on this issue further in Section 4.5.

3.5 Putting Things Together

Similarly as correspondence realism can stand in the way of an active view of knowledge, a certain kind of reductionism can stand in the way of the kind of pragmatist and pluralist metaphysics advocated in this chapter. This species of ontological reductionism, which I will call **Legoism**, considers any objects to be put together by a simple assembly of unchangeable units in the manner of building things with Lego bricks. Legoism is closely linked to the mereological and set-theoretic habits of philosophical thinking discussed in Section 2.2. Legoist metaphysics provides a hostile climate for pragmatist metaphysics because the assumption of unalterable basic building-blocks of matter encourages the fallacy of pre-figuration. And even though in principle there could be multiple sets of immutable basic building-blocks, that becomes difficult to maintain if the building-blocks are imagined to be not mind-framed. Legoism is often assumed to be supported by science; however, a careful look at modern physics shows that physical combination is not Lego-like assembly. It is also not the case that Legoist intuitions originate from our everyday experiences. Rather, they are instilled by the mental habits of making Legoist analysis.

Against Legoism

Before I leave behind the explicit discussion of ontology, I must address a certain well-entrenched metaphysical doctrine that stands powerfully in the way of the acceptance of pragmatist metaphysics. In a light-hearted terminological move I am going to call it 'Legoism', because it pictures everything in the world as composed of unchangeable units, like Lego bricks. According to Legoism, everything is made up by a simple assembly of unchanging parts, and can be decomposed cleanly into those parts. This view seems to be very much a part of our philosophical and scientific common sense.

How exactly is it that Legoism stands in the way of pragmatist metaphysics? The logical connection is not tight, but there is a strong intuitive push. Legoist metaphysics provides a hostile climate for pragmatist metaphysics, even though it does not directly contradict it. This is because Legoism often serves as an excellent vehicle for the fallacy of pre-figuration. Legoists usually assume that the unchanging fundamental units of their analysis are mind-unframed parts of Reality, even though this is strictly speaking not required by Legoism itself. In principle it would be possible to combine Legoist thinking about composition with a pluralist allowance of multiple sets of fundamental units, but this possibility is usually not entertained. Any other valid ontology at a higher or more complex level of existence is assumed to be reducible to the fundamental building-blocks.

Thus, the usual version of Legoism holds that everything can be ultimately broken down to the one-and-only set of fundamental building-blocks that cleanly make up everything else. It would seem that everyone accepts something like the hierarchy of ontological composition as given by Paul Oppenheim and Hilary Putnam in their classic paper on reductionism (1958, p. 9): (1) elementary particles; (2) atoms; (3) molecules; (4) cells; (5) (multicellular) living things; (6) social groups. Dupré has given a trenchant critique of 'microreductionism', which he defines as 'the view that the ultimate scientific understanding of a range of phenomena is to be gained exclusively from looking at the constituents of those phenomena and their properties' (Dupré 1993, p. 88). Initially I followed a key line of argument from Dupré: even if we accept ontological microreduction, epistemic microreduction does not follow. Now I think that it is also important to subject ontological microreductionism to full critical scrutiny. We should not concede so readily that composition is Legoist, that a material whole is just a juxtaposition of its parts. Once it is agreed that there are elementary building-blocks of nature 'out there' independently of all conceptualization, it is easy to argue that all objects are *obviously* mereological sums of those building-blocks.

Modern Physics vs. Legoist Metaphysics

Legoism is allegedly backed up by modern physical science, but it is not. James Ladyman, Don Ross and David Spurrett make this point very strongly as part of their argument against a priori metaphysics shaped by outdated scientific common sense (Ladyman and Ross 2007, ch. 1). I endorse this aspect of their critique, and would add that showing the limitations of Legoism does not even require cutting-edge contemporary physics. Attention to the actual practices of physical composition and decomposition over the long history of the physical sciences will show that parts are not all there is to a whole. 'Parts' are only the salvaged remnants of a whole that has been *shattered*, physically or conceptually. This is easy to see if we think about social situations: if we destroy society completely, will the individuals left be the same beings that you recognize in a functioning society? I want to argue that the situation is actually very similar with physical bodies. If we pay attention to successful analytic and synthetic practices in chemistry and physics, we will see that they do not go in Legoist ways.

Legoism is closely related to what I have called **compositionism** in chemistry, defined as 'the notion that chemical substances are made up of stable units that persist through chemical reactions' (Chang 2017e, p. 218; also Chang 2011b; 2012a, ch. 1). Legoism is generalized compositionism, extended beyond chemistry. I could have called it 'atomism', but I have avoided that term because what we call 'atoms' in modern science are breakable and changeable – not very 'atomistic'! Legoist thinking gained ascendancy in chemistry and physics during the eighteenth and nineteenth centuries. It had to displace well-entrenched metaphysical alternatives, especially neo-Aristotelian hylomorphism, which conceived specific materials as the outcome of the imposition of form on substance. By the early modern period scientists began trying to decompose things into their constituent parts, but there were worries that the alleged processes of decomposition might be altering the substances being analysed. For example, the application of strong heat was commonly thought to break things down, but the cogency of 'fire-analysis' was questioned by the likes of Robert Boyle (see Debus 1967): how could one be sure that the application of fire was merely breaking things up into their constituents, rather than altering their very nature, or at least getting fire-particles sticking to them? Similar doubts were also raised concerning other analytical methods, such as the dissolution of substances by the application of acids.

That was all in the prehistory of respectable science, you might say. Let us, then, come into the twentieth century, and take a look at the 'atom-smashing' practices of modern experimental physics. Atom-smashing has never been Lego-like disassembly: when atomic nuclei are broken up, energy is almost always added or subtracted; given the interconversion of mass and energy, this means that the amount of matter is not preserved in decomposition. It won't do to suggest that the Lego-like picture is *approximately* true: no theory that has to dismiss nuclear bombs as unimportant details should be regarded as 'approximately true'. As empirical evidence for mass–energy equivalence ($E = mc^2$), what is invoked most often are particle-collision experiments in which the masses of the ingredients do not add up exactly to the sum of the masses of the products (see Fernflores 2012, esp. sec. 4). Especially famous is the 1932 experiment of John Cockcroft and Ernest Walton, who bombarded a lithium nucleus with a proton and obtained two helium nuclei (α-particles). Their measurements showed that the sum of the masses of the reactants was $1.0072 + 7.0104 = 8.0176$ amu,[19] but the masses of the products only added up to 8.0022 amu, indicating that 0.0154 amu had 'disappeared', turning into other forms of energy. Thus Lavoisier's principle of the conservation of mass was overturned in physical and chemical practice, after over a century of dominance. An inspection of the periodic table of elements easily shows that the masses of atoms are slightly different from the sum of the pre-combination masses of protons and neutrons (and electrons) that constitute them. These have been regarded as indisputable facts for many decades now by physicists and chemists, but the basic metaphysical implications of such facts have not got through to the sensibilities of philosophical reductionists. If we still regard mass as the primary indicator of the amount of matter, then it is clear that the amount of matter is not preserved in elementary-particle collisions or nuclear reactions. That does not present a

---

[19] 1 amu (atomic mass unit) is 1/12 of the mass of carbon-12 in its ground state.

problem for the conservation of *energy*, of course, but it does destroy the naïve notion that atoms are simply put together from elementary particles with fixed masses. Atomic nuclei are not mereological sums of protons and neutrons; they are not made up of protons and neutrons in a straightforward Legoist sense.

Experiments in high-energy physics do not support the naïve view of elementary particles as unchangeable building-blocks of matter. When two protons collide with each other in a particle accelerator, a whole host of other particles are created:[20] should we say that a proton (or two protons together, somehow) already *contained* these particles? And should the phenomena of pair-creation and pair-annihilation lead us to conclude that a pair of photons consists of an electron and a positron, or vice versa?[21] And when a photon is absorbed by an atom, it ceases its existence but raises the energy level of the atom; so the photon is not an unchangeable unit, not even a persisting one. These are merely a handful of illustrative examples. Generally speaking, in the physics of so-called 'elementary' particles, smashed-up pieces do not necessarily pre-exist in the whole. This recognition led Geoffrey Chew to advance his 'bootstrapping' view of elementary particles, according to which elementary particles are made up of one another.[22] This view was sidelined with the advent of quarks and the Standard Model, but it may be worth revisiting, after all. That is also to say, the early modern doubts about physical mereology have returned with a vengeance. Again, I am only invoking very basic experimental facts here. I am not even entering into the difficult ontological questions raised by quantum superposition and entanglement, or the indistinguishability of identical particles, or virtual particles and vacuum fluctuations, or quark confinement, all of which are bound to complicate the picture much further and in all likelihood in anti-Legoist directions.

How Philosophy Should Resist Legoism

At this point some philosophers may say: 'But what these scientific experiments apparently seem to show can't really be the true metaphysical picture. There *must be* unchangeable basic building blocks, and everything must

---

[20] See a friendly presentation of the basic facts and ideas in the 'International Physics Masterclasses' section on proton collisions (https://atlas.physicsmasterclasses.org/en/zpath_protoncollisions.htm).

[21] Positron–electron pair-annihilation is now even in the realm of familiar technology as the basis of PET (positron emission tomography) scans, which work by the injection of positron-producing radioactive atoms into the body; it is not something we can afford to ignore in our thinking about the world. See Shang (2021) for an illuminating historical and philosophical account of PET.

[22] For more on the metaphysics of bootstrapping and the S-matrix theory, see McKenzie (2011).

ultimately be made of them.' Of course, no amount of scientific knowledge can *prove* that the metaphysical reality is *not* like Lego. I just think that Legoism is not a universally productive way of thinking about physical reality. If we look at the scientific situation closely we will see that Legoism has been most successful not at the ultimate level of micro-reality, but at an interesting middle-level that is the realm of molecules and ions (see Chang 2017d). This certainly does not inspire metaphysical Legoism about the ultimate constitution of matter.

But where *do* people get the intuition that reality must be Lego-like? If Legoism is not supported by successful practices in physical science, then where do the widespread intuitions in its favour come from? It could be that the compositionist intuitions are rooted in our everyday life. Robert Northcott, in a serious joke, observes that we must have Legoist intuitions because we all grew up playing with Lego. Since Lego itself is a twentieth-century invention it can't have been responsible for the advent of compositionism in premodern science, but could it be that much of our everyday life is like playing with Lego? We can smash a plate and glue it back together, build a house out of bricks, and take apart a watch and put it back together. But these quotidian practices of composition actually do not work in Lego-like ways. 'Medium-sized dry goods' generally do not stick to each other. You can't build a brick wall without mortar, or glue things together without glue. How glue works is not at all like Lego (and actually quite mysterious!), so our experience of gluing (or stapling, or clamping, or strapping) things together does not explain why we have Legoist intuitions. In biology, too, a multicellular organism is not just a bunch of cells put next to each other; the intercellular matrix helps cells hold together.[23] And a social community cannot be built without individuals undergoing changes through their mutual associations, changes that enable the associations in the first place. Lego, first marketed as '*automatic* binding blocks', was such a commercial and cultural success precisely because it was a very clever and unusual arrangement (an ingenious combination of rigidity, elasticity and friction), in which bricks do stick together without the help of anything else![24] Almost nothing else in nature or human life behaves like Lego,[25] and that is the secret of its success. We do not live in Legoland.

I suggest that that our common Legoist intuitions come not from practical experience, but from a quasi-Kantian conceptual necessity of the kind

[23] I thank Matt Meizlish for teaching me about the intercellular matrix.
[24] For the history of Lego as told by the company itself, see www.lego.com/en-us/aboutus/lego-group/the-lego-group-history (last accessed 13 September 2021).
[25] Gretchen Siglar points out that Velcro is one rare example.

that I discussed above in Section 3.2. Legoist intuitions are ontological principles that are necessitated when we choose to carry out Legoist analysis, namely the activity of understanding an object as a mereological sum of its parts, whose identity or essential nature is not affected by any combinations into which they enter.[26] So, if we do choose to carry out a Legoist analysis, then of course it makes no sense not to adopt Legoist ontology; that would render our activity incoherent and unintelligible. And if we do routinely carry out Legoist analysis, we may understandably form Legoist habits of mind. We often do engage in such analyses in certain activities of everyday life. Simple-minded accounting is a very good example; perhaps most fundamentally, the standard kind of arithmetic that we learn in childhood is firmly founded on Legoist intuitions. We also engage in Legoist analysis in theoretical science, for example when we apply various conservation laws. However, this is not to say that we should always engage in Legoist analysis. It is futile to do so where we cannot find reliably persistent units out of which the objects of our interest can be said to be made. Whether or not there are such units is an empirical question, a contingent matter, which can only be clarified in a pragmatic way, by devising and attempting to carry out actual operations of physical assembly and disassembly.

The main source of widespread intuitions in favour of ontological reductionism is not successful scientific practice, but our *predilection* for Legoist analysis. It is unwise to let ourselves be guided by these intuitions in situations in which attempts at Legoist analysis deliver little empirical success. Mereology and set theory may apply nicely to some aspects of reality, but that is a contingent matter. It should not simply be presumed that these schemes apply to realities. My pragmatist view is that any conceptual scheme should prove its worth by producing beneficial results of some kind. Judging from the findings of modern chemistry and physics, it seems that the mereological part–whole relation is an inappropriate framework for understanding actual physical combination. Physical fusion and disintegration may violate axioms that one considers reasonable or even indispensable in mereology, such as transitivity. If so, the most reasonable conclusion may be that physical composition is *not* a matter of part–whole relation in standard mereology. And if Legoism doesn't work in chemistry and physics, then it is not likely to work in other sciences. This destroys the foundations of the grand microreductionist

---

[26] Reductionist scientists and philosophers often add two further ideas, which are actually not necessary for Legoist analysis per se: (1) that there are fundamental parts that cannot be further decomposed, and (2) that there are only a small number of types of fundamental parts or units.

strategy. We need to make sure that quasi-Kantian conditional necessity does not degenerate into pseudo-Kantian metaphysical prejudice.

These thoughts about Legoism also offer an insight into how to do naturalistic metaphysics. Naturalism should not mean just following any metaphysical consensus that scientists reach among themselves, or subscribing to any metaphysical pictures implied by the best scientific theories of the day. The historical development of science shows plenty of instability and contingency. If we follow the 'verdict of science' blindly, we risk mistaking scientists' presuppositions as warranted conclusions of inquiry. Rather, naturalism should oblige us to adopt hard-earned insights coming from well-established scientific practices, but only after a thorough philosophical analysis. Attention to the actual practices of chemistry and physics reveals that there has never been unequivocal scientific warrant for Legoism. Nineteenth-century structural chemistry seems to have been a bright blip in the history of Legoism in science, not the moment of its firm establishment for all future science.

CHAPTER 4

Truth

4.1 Overview

Why Worry about the Concept of Truth?

If you are not a professional philosopher, or even if you are one, why should you concern yourself with the theory of truth? To practical people it may seem that they know very well how to tell what is true and what is not true; there is no need for philosophical disquisition. But truth is not such a straightforward matter, and any complacency you might have felt about it should have disappeared in this dystopian start to the twenty-first century, with many societies utterly unable to reach consensus on some very basic matters of truth. The liberal faith that we can all agree on the facts and have a polite discussion over opinions and values is being seriously challenged by people not accepting each other's facts.

For a telling sign, witness how the *New York Times* marketed itself in the crucial election year of 2020 in the United States: 'Life needs truth. The truth is essential. *The New York Times* – Subscribe now.'[1] But how do we evaluate the claim that *this* newspaper will tell you the truth? Easy enough to trust that the *New York Times* does not deliberately print lies, but honesty is not enough, since misinformation is so often conveyed very earnestly. And it is not good enough to say that it employs stringent procedures that are designed to filter out misinformation. In order to know if such procedures work, we already need to know that the specific pieces of information they filter out are falsities and what they allow in are truths – which brings us back to square one. We are hit with the same sort of difficulty when we try to say everyone should trust science to tell us the truth.

---

[1] This statement occurred at the end of a video advertisement to potential subscribers: www.nytimes.com/subscription/truth/truth-is-essential (last accessed 17 October 2020).

So there is urgent need to think more carefully about what truth is, and how we can know it. Unfortunately, some of the most thoughtful cutting-edge academic work in the humanities originating in the late twentieth century has had the adverse effect of casting destructive doubt on the traditional ideas of truth, fact, objectivity and rationality without offering anything convincing to take their place. There have been attempts to enter into thoughtful discussions of the 'post-truth' phenomenon (e.g., Sismondo 2017; McIntyre 2018), but philosophical thinking on this issue tends to be hampered by an attachment to the non-operational idea of truth as correspondence to the ultimately inaccessible reality (see Chapter 2). We need to rethink the notion of truth itself carefully, in order to make an effective defence of truth without making indefensible claims or putting up ideals with no bearings on actual practices.

With that task in mind, it is particularly disappointing to see that among the most rigorous philosophical thinkers a notable current tendency is *deflationism* about truth, which tries to avoid vexing questions about the nature of truth by turning it into a maximally empty concept. Deflationists take the function of the word 'true' as simple affirmation. Daniel Stoljar and Nic Damnjanovic (2014) open their article on the subject thus: 'According to the deflationary theory of truth, to assert that a statement is true is just to assert the statement itself.' Stating that 'Proposition P is true' only amounts to the assertion of P. A simple-minded version of Tarski's disquotation schema is repeated endlessly in this context: 'Snow is white' is true if and only if snow is white.[2] That is all there is to the meaning of 'true', the deflationists argue. Or as Paul Horwich puts it (1998a, pp. 103–4): 'The basic thesis of deflationism, as I see it, is that the equivalence schema "The proposition *that p* is true iff *p*" is conceptually fundamental.'[3]

Deflationism is not objectionable in itself, but for anyone trying to get a sense of what it means for certain propositions to be true in concrete situations, it is not a useful doctrine.[4] Especially in the empirical domain, it is crucial not to restrict the function of the truth concept to simple

[2] As Greg Ray (2018, p. 699) notes, this is done without mentioning the fact that not all 'T-sentences' need to be true, and that Tarski himself wrestled with the liar paradox.

[3] Horwich explains further: 'By this I mean that we accept its instances in the absence of supporting argument: more specifically, without deriving them from any reductive premise of the form "For every x: x is true = x is such-and-such" which characterizes traditional ("inflationary") accounts of truth.'

[4] In this judgement I broadly follow Cheryl Misak's (2007b) pragmatist critique of various deflationary accounts of truth. However, I hesitate to follow the Peircian conception of truth advocated by Misak.

affirmation. As Price (2003) argues, a key function of the truth concept is to prompt people to engage in debates aimed at resolving disagreement. If I ask you 'Is this statement really true?' I am quite likely demanding to know whether there are good enough grounds for it. If we are asked 'Is it true that putting more carbon dioxide in the atmosphere raises the global mean temperature?', I think the questioner would probably be expecting an answer like 'It must be the case because carbon dioxide absorbs infrared radiation very well, trapping the heat escaping from the earth into space in that form', or 'At least there seems to be a clear correlation between higher carbon dioxide levels and average global temperature.' If instead we responded to this demand in a deflationist way by simply reasserting the statement, we would be missing the point of the question. In the empirical domain we seek truth that is learned and tested by experience, rather than truth that is a matter of mere assertion, self-consistency and logical inference.

Deflationists may be quite correct that 'philosophers looking for the nature of truth are bound to be frustrated ... because they are looking for something that isn't there' (Stoljar and Damnjanovic 2014). But I think deflationists are wrong in their diagnosis of where the trouble for 'inflationism' lies. It's not that there isn't *anything* substantive to truth, but only that there are *many* things meant by truth, so that attempts to find a *universal* notion more substantive than the deflationary one fail.[5] In my view, more productive than deflationism is 'truth pluralism' as advocated by Michael P. Lynch and Nikolaj Pedersen: 'There is thus a range of properties (correspondence, superassertibility, coherence, etc.) that constitute truth for different domains of discourse.'[6] In Section 4.2 I will give a more considered view of truth pluralism, also building on the homespun version of it that I have given before (Chang 2012a, sec. 4.3.1) without realizing that there was already considerable literature on it.

Primary and Secondary Truth

In thinking about the different meanings and functions of truth, it is helpful to start with a distinction between what I will call **primary truth** and **secondary truth**. A true proposition is true in the secondary sense if its truth derives from the truth of other propositions (or we might say that

---

[5] Kevin Scharp (2013) argues that even deflationary truth needs to be separated into two notions, in order to avoid contradictions.

[6] Pedersen and Lynch (2018, p. 546); see Lynch (2009) for further discussion.

it possesses secondary truth, or that it is a secondary truth). A true proposition whose truth does not derive from the truth of other propositions is true in the primary sense (or, it possesses primary truth, or it is a primary truth).[7] The expression 'derives from' is a loose way of speaking, and gets at the sense of grounding. Rather than going deeply into the nature of grounding, I would like to think more practically about specific ways in which the truth of one proposition depends on the truth of other propositions. Perhaps the most straightforward way in which secondary truth can be constituted is by enumerative induction in a finite set. 'All of my cats are black' is true in virtue of it being true that each and every cat of mine is black. Similarly straightforward would be grounding by deductive consequence from other propositions that are true. Various other ways of grounding secondary truth in primary truth will enter the discussion later on.

One distinction that needs to be made carefully at the outset is between the constitution of truth and the justification of belief. I want to focus on the question of what makes something true, not the epistemic conditions concerning how we can *know* what is true and what is not. If justification means finding good reasons for believing something, then the line of justification may or may not follow the line of constitution of truth. You may justifiably believe that all of my cats are black because Stuart told you so and Stuart is generally a reliable witness, but the truth of the proposition 'All of my cats are black' does not actually depend on Stuart being a reliable witness (or on your wifi connection being good enough to allow you to hear clearly what he said). Now, sometimes the line of justification does follow the constitution of truth – if, for instance, you justified your belief by seeking out each of my cats and confirming their blackness in each case.

For our present purposes, it is helpful to set aside the question of justification. The constitution of truth does not suffer from vicious circularity or infinite regress. The primary–secondary distinction concerning truth is clearly a hierarchical one, and we need to recognize that the constitution of primary truth is a fundamentally different kind of thing from the constitution of secondary truth. But this is a tricky process, as Wittgenstein memorably indicated in *On Certainty*: 'It is so difficult to find the *beginning*. Or, better: it is difficult to begin at the beginning. And

---

[7] My terminology harbours a deliberate ambiguity that I think is convenient, and harmless enough: 'truth' means both the quality of being true, and a proposition (or statement) that is true. This is similar to 'reality' meaning both the quality of being real, and an entity that is real.

not try to go further back.'[8] I will make some further clarifications on the interactions between primary and secondary truth in Section 4.2, but I think I have now said enough to allow me to get on to the main idea I want to advance in this chapter.

Truth-by-Operational-Coherence

I want to craft a non-deflationary theory of truth suitable for **empirical domains**, including science and much of daily life as well. In an empirical domain we find realities (as defined in Chapter 3), entities that do not do as we wish, which are mind-framed but not mind-controlled. In an empirical domain we learn facts through experience; it is a different kind of setting from the a priori domains of logic and mathematics as they are commonly understood, where truth just follows from postulates that are adopted by the mind. It is also different from fictional domains, where truth can be imagined as we wish. Whether moral or religious truths are in empirical domains is a controversial question, which I will not try to answer here. But I do want to offer a conception of truth that can be usefully applied in whichever domains that one may treat as empirical.

The key task is to understand what constitutes primary truth in empirical domains, and this is my proposal: *a statement is true to the extent that there are operationally coherent activities that can be performed by relying on its content*. In parallel to that, taking a proposition as the content of a statement, we can say: *a proposition is true to the extent that there are operationally coherent activities that can be performed by relying on it*.[9] Let me call this **truth-by-operational-coherence**. The notion of 'operational coherence' here is what I developed initially in Sections 1.1 and 1.4, and already used in my characterization of reality in Chapter 3. I hasten to add that I am by no means presenting truth-by-operational-coherence as the only kind of truth there is. As I will explain fully in Section 4.2, I subscribe to pluralism concerning truth: there are many different notions of truth, which have different uses in various domains and contexts. But I do want to propose that truth-by-operational-coherence is what constitutes primary truth in empirical domains, therefore something that we should centrally concern ourselves with in the philosophy of science.

---

[8] 'Es ist so schwer den *Anfang* zu finden. Oder besser: Es ist schwer am Anfang anzufangen. Und nicht versuchen weiter zurück zu gehen' (Wittgenstein 1969, p. 62, §471, emphasis original).

[9] I reject the idea that statements do not bear truth-value, as it does excessive violence to normal language.

Now, let's consider more carefully how operational coherence, which is a property of an activity, relates to truth, which is a property of a proposition or a statement. In order to build up some initial intuitions, let us take a fresh look at the kind of empirical statements that we do take to be so secure that they can serve as the ground for other true statements, and ask why we regard them to be true:

'The ground is firm.'
'Here is a hand.' (G. E. Moore)
'The people that I see when I walk into a room are really there.'
'When I wake up in the morning, the earth will still be here.'

Why do we take such propositions as unquestioned truths? It is not because they are justified by some other propositions that are more fundamental or more secure. This is where we need to remember Wittgenstein's warning against trying to go back further than the beginning. In my own idiom: we should not mistake the above propositions as secondary truths. They are primary truths, which means that the grounds for the truth of these propositions need to be found in themselves (or rather, in their functions). So my proposal is that they are true-by-operational-coherence, true in the sense that numerous activities that we routinely carry out are reliant on them.

Take the proposition that the ground is firm. We carry out most activities of our earthly lives on the basis of this assumption, though freak events do sometimes happen to disturb the assumption, for example when sinkholes suddenly open up. I remember experiencing a major earthquake (and various aftershocks following the big one) in October 1989 in Palo Alto, near San Francisco. (I was just starting my graduate work in philosophy, in fact sitting in a colloquium when the tremors started, so my philosophizing will never be entirely free from that experience.) For a short while afterwards all aspects of life were different, not even being able to walk around assuming that the ground was fixed. This strange existence did stop after a while, but that would have been different if major earthquakes had kept happening every few days. But as long as the reliance on the statement in question supports an effective way of life, then we go on regarding it as true. And why shouldn't we say that it *is* true that the ground is fixed (even though it is also true that the earth is moving through space incredibly fast), when and where our activities premised on that idea are operationally coherent?

In science, too, the kind of truth possessed by foundational propositions is truth-by-operational-coherence. Why should we take direct empirical observations as true in general, even though philosophical sceptics have shown plenty of reasons for doubting them? Because it turns out that we can carry out a great number of operationally coherent activities on the basis of taking our own and others' sincere observational reports at face value (reserving doubt for particular and unusual circumstances). The same also goes for basic propositions of a more theoretical nature – for example, that the speed of light is constant in all directions regardless of the motions of its source or receiver, or that all genetic information is contained in the DNA molecule. Even though these basic theoretical propositions are often taken as axioms, that is, held true by decree, continued adherence to the axioms would be pointless unless the activities that they support were operationally coherent. This is the spirit of the arguments by Dewey and C. I. Lewis that a priori elements, even the axioms of logic, can in the end only be justified pragmatically by the operational coherence of the reasoning activities that they support (see Sections 1.6 and 3.2). Certainly in the realm of empirical science, theoretical propositions cannot be upheld as true *simply* by decree. Even though some central empirical propositions are treated dogmatically (like those in the 'hard core' of a research programme in Imre Lakatos's view of science), that should only be a temporary and provisional situation. We try to devise epistemic activities on the basis of the postulates, to see how well we can come up with coherent activities.

As I will explain further in Section 4.6, truth-by-operational-coherence is a robustly pragmatist notion – not only in paying attention to practical consequences of our beliefs, but in having our thoughts fully rooted in experience. In relation to the standard critique of the classical pragmatist theory of truth, there is one point worth stressing immediately, and briefly for now: the definition I offer here is precisely *not* in the spirit of 'something is true if it is convenient for me to believe'. A coherent activity is a difficult thing to devise, and it will only work out if our assumptions entering into it are suitable; that is how empirical statements must be put through the test of experience. Operational coherence carries within it the constraint by nature, which gives truth-by-operational-coherence the mark of mind-independence in the sense of the absence of mind-control, which is something that many realists rightly value in their favoured correspondence notion of truth. The basic pragmatist intuition is that primary truth is something that we can *live by*. Unlike religious truth in its ordinary conception, which one also lives by, empirical primary truth is tested by experience; it is revisable in response to the expansion of our experience or changes in our situation. When we conceive truth in terms of coherent activities, it is squarely placed in the realm of active knowledge as I presented it in Chapter 1.

If we understand empirical primary truth as truth-by-operational-coherence, what about empirical secondary truth? Here we have a natural way of rehabilitating the correspondence notion of truth, in a more down-to-earth form, along the lines that I suggested in Section 2.5. I think it would be uncontroversial that the substantive theory of truth most suitable for empirical secondary truth is the correspondence theory. One would have to work out exactly how secondary truth is constituted from primary truth in empirical domains, going beyond deduction and enumerative induction as discussed above, but that is a tractable problem that can be tackled plausibly. When we see the nature of truth-by-correspondence as secondary truth, there is no need to imagine a 'transcendent' type of reality to which our theories need to correspond in order to possess truth-by-correspondence.

Again, it is important not to confuse the constitution of truth itself and the justification of our knowledge of it. The operational coherence of activities relying on a true proposition are not *consequences* or *indications* of its truth, from which we may infer the truth. Rather, operational coherence is *constitutive* of truth. As Dewey put it ([1907] 1977, pp. 68–9): 'the effective working of an idea and its truth are one and the same thing – this working being neither the cause nor the evidence of truth but its nature'. As with the notion of reality in Chapter 3, what I am making here concerning truth is a semantic move, in the sense that I am proposing what we ought to *mean* by 'truth', in order to render it as a useful concept.

Against Truth-Absolutism

Truth-by-operational-coherence is a robust enough notion to serve all the main functions that we would want a concept of truth to serve in empirical domains. And it can serve those functions without being an absolutist notion. For example, we should say that Newtonian mechanics remains quite true (true-by-operational-coherence) in its proper domain. It would be not only annoying and pedantic, but actually unwise, to insist that we now know that Newtonian mechanics is false and that it just gives approximately correct empirical predictions in certain situations. If we do that, by the same lights we would be obliged to say the same about general relativity, quantum mechanics, or any other theory we now have. Then 'true' would become a designation that can never actually be used.

Should we really want to discard or *incapacitate* such a crucial concept for our intelligent life? That would be like insisting that we should never say things like 'She is a good person' unless someone is unfailingly and perfectly good. Should we be saying that she is an 'approximately good' person, or a 'benisimilar' person? No, better to stick with the common sense of saying that she is a good person, but like all of us she could become still better in various ways. Many physicists have now become content to say that every theory is an 'effective theory', true with limited scope (see Cao and Schweber 1993); their main focus is on the pertinent energy level at which each theory functions, but the lesson can be generalized to cover other parameters, too.

Generally speaking, truth-by-operational-coherence is a *qualitative* attribute. A proposition is true to a higher degree if it supports a larger number and variety of operationally coherent activities, and if such activities rely on the proposition in question more strongly; moreover, operational coherence itself is a matter of degrees. And it is the quality of truth itself that is a matter of degrees, not just our knowledge of it or our belief in it. This is actually consonant with everyday usage: as Austin noted ([1950] 1979, pp. 117, 130–1), 'very true', 'true enough', etc. are perfectly sensible locutions, and it is unreasonable to try to reduce ordinary judgements of truth to yes/no. The spirit of Austin's observation has now become current again, thanks to its revival by Catherine Elgin in her *True Enough* (2017). And the degree of truth I am speaking of will not be quantifiable in a simple numerical way, nor reducible to probabilities. In practice, truth-in-degrees is already a widely accepted notion in the philosophy of science. Many philosophers, most of all in the course of defending standard scientific realism, have already fallen into the habit of speaking about 'approximate truth', and Richard Boyd (1990) has argued convincingly that it is not possible to maintain scientific realism without relying on some notion of approximate truth. Perhaps one could say that 'approximate truth' is an imprecise way of speaking, and what we are really talking about is an *approximation* to the truth, while truth itself remains a yes-or-no matter. But I do not see what would be gained in preserving binarity for truth in that way. On the contrary, as hinted in Section 1.6, the current flourishing of many-valued logic with its use in computing and artificial intelligence would tend to suggest that departing from a binary notion of truth is a legitimate and potentially very useful move.

Another factor that makes truth non-absolute is its finite scope, or domain-specificity. According to my definition of truth-by-operational coherence, a statement that is true in a certain domain can easily fail to be true in other domains (i.e., it may not support coherent activities there). And who would deny, on reflection, that a given statement can be true in some domains and not in others? We are all familiar with situations where a law of nature is only true in some cases, such as the classical laws of motion being quite wildly invalid in various quantum-mechanical domains.[10] It would be useful for us to get into the *habit* of always asking 'where/when is this statement true?' as an antidote to absolutist and universalist tendencies. Now, many people will be uneasy about attaching limited scopes to truth. This leads to what I might call the 'witchcraft objection': a patently untrue theory (e.g., of witchcraft) may work well enough (support coherent activities) within a small domain, but it seems wrong to say that it is *true* even within that domain. This issue will be dealt with fully in Section 4.4, but for now let me just express my view that what we are dealing with here is an unnecessary fear. There are reasons for which we should not want to grant truth to the theory of witchcraft even in a narrow domain (and similarly for climate change denial, vaccine refusal, 'young-earth' creationism, flat-earth cosmology, and so on): it does *not* in fact support operationally coherent activities very well in *any* domain; the claim that it worked well in certain situations is wildly exaggerated. On the other hand, if we do have a theory that actually works well and does not contradict other established truths within a given domain, there is nothing wrong with regarding such a theory as an empirical truth in that domain. I think it is correct to say things like: 'Newtonian mechanics remains true in its domain of application.' And it has not shown to be less true in most of its old domain than quantum mechanics and general relativity are, because no one has yet tried applying the latter theories (unmixed with classical mechanics) to most classical situations.

Yet another non-absolute aspect of truth-by-operational-coherence is plurality. If truth-by-operational-coherence only makes sense in the context of an activity (or within a whole system of practice), then there can be different sets of truths belonging to mutually incommensurable activities or systems. This points to epistemic and ontological pluralism, as a separate issue from 'truth pluralism', which is about the meaning of 'truth' itself. The plurality I have in mind now is all exhibited under one concept of truth, namely truth-by-operational-coherence. It is *true* that light is an

---

[10] One could avoid the domain-specificity of truth by writing the domain-restriction explicitly into the statement in question and treating the restricted statement as strictly true or not. That would be a losing game because not all possible domain-restrictions can be anticipated and specified.

electromagnetic wave, and it is also *true* that light is composed of photons. This feature of truth-by-operational-coherence is very consonant with the pluralism concerning science articulated in my earlier work, which is 'the doctrine advocating the cultivation of multiple systems of practice in any given field of science' (Chang 2012a, p. 260). I will discuss pluralism further in Section 4.5, and again in relation to realism in Chapter 5. Recognizing the simultaneous truth of statements made in different systems of practice does not easily result in a logical contradiction, even when they seem to say conflicting things. This is for two reasons. In many cases the apparently contradictory statements are not employed in the same situations, so they do not actually clash. And when they do concern the same situations, there is usually a sufficient degree of semantic incommensurability so that they do not directly contradict each other, talking past each other instead. It is important to note that the multiplicity and non-absoluteness of truth do not amount to a crude or extreme form of relativism, or 'idiot relativism' as Arthur Fine (2007, p. 64) calls it. Within each system of practice, truth-by-operational-coherence is fixed in a way that is not controlled by our wishes or expectations. There is no reason to fear the multiplicity of such mind-framed yet mind-uncontrolled truth.

In the remainder of this chapter I will give further elaborations on my perspective on truth. As a preliminary step, in Section 4.2 I will present various other valid conceptions of 'truth' differing from truth-by-operational-coherence, and discuss what kinds of situations render these notions suitable. In Section 4.3 I will give a fuller discussion of truth-by-operational-coherence, and elaborate further on how this notion of truth is not an absolutist one, and try to anticipate worries that people may have concerning this lack of absoluteness. Section 4.4 will argue that we should have a concept of empirical truth according to which truth comes attached with a specific scope, and Section 4.5 will come to terms with the fact that multiple truths-by-operational-coherence can co-exist in a given subject area. Finally in Section 4.6 I will link this notion of truth back to the traditional pragmatist ideas about truth, and clear up some persistent misunderstandings about the classical pragmatist ideas.

4.2   Different Kinds of Truth

Before I go on to discuss my notion of truth-by-operational-coherence in detail, it will be useful to see it in relation to various other meanings of truth. Following the general spirit of pragmatism, and in sympathy with 'truth functionalism', I identify and distinguish various meanings of truth according to the different types of activity within which the terms 'true' and 'truth' acquire their meanings, focusing on empirical science and the most common practices of daily life. This analysis will produce partially overlapping meanings of truth, rather than a disjunctive set of mutually exclusive truth-concepts. Truth-by-assertion is based on a minimalist idea that asserting a proposition is equivalent to saying that it is true. We have truth-by-honesty when someone tells you without distortion what e thinks or feels. Truth-by-decree obtains when we postulate something to be true. None of these senses of truth captures what we mean when we say that something is empirically true, which points to truth-by-operational-coherence. And as discussed in Section 4.1, there is also secondary truth (which I will also call truth-by-comparison), which is a matter of agreement with other propositions that are already established as true. It is important to note that various types of truth often operate in conjunction with one another, which creates interesting complications in using the notion of truth in real practices.

Functionalism and Pluralism Concerning Truth

What should a philosophical theory of truth do? I think it should give a synoptic view elucidating how the concept of truth is used in various domains. This is according to my late-Wittgensteinian inclination to take meaning as arising from use, treating truth in this regard like any other term. My debt to Austin's perspective, in general and specifically on truth, has already been noted. Within the 'neo-pragmatist' tradition Robert Brandom (1994) and Huw Price (1988; 2003) have stressed the need to consider functions of the concept of truth. More recently Lynch (2001; 2009) has added to the advocacy of 'truth functionalism'. There are subtle differences among these various positions and perspectives, especially concerning whether we are inquiring into the *meaning* or *function* of truth. Instead of entering into a debate about the differences, I will build on what I see as a common core: attention to how the concept of truth is actually used in various activities. So the task here is not to review various theories of truth in order to say which is the correct one, but to see how different conceptions of truth are operative in different contexts.[11] And this approach is also consonant with a cross-cultural comparative approach to the notion of truth, as advocated and practised for example by Alexus McLeod (2016).

---

[11] For general surveys of various theories of truth, see Kirkham (1992) and Glanzberg (2018b).

Table 4.1. *Various concepts of truth*

Concept	Core intuition	Associations
Truth-by-assertion	'P is true' means 'P'.	Deflationism, disquotation
Truth-by-honesty	P is true if it is asserted sincerely.	Truth-telling, witnessing
Truth-by-decree	P is true if we decide to affirm it.	Postulation, axiomatic systems
Truth-by-operational-coherence	P is true if it facilitates coherent activities.	Pragmatism, embedding in active knowledge
Truth-by-comparison (secondary truth)	P is true if it agrees with other truths already established.	Correspondence, confirmation

It is easy to see that the term 'truth' or 'true' is used in many different ways in practice. In Lynch's wry expression (2001, p. 723): 'The history of attempts to identify the property that all and only true propositions have in common has not been a happy one.' This has led some philosophers to adopt a pluralism concerning the concept of truth.[12] Crispin Wright notes that in certain domains truth is based on a correspondence-like relation, while in others it is a matter of assertibility. Pedersen and Lynch (2018, p. 546) draw a very instructive analogy with the concept of 'winning': there is 'variation in terms of what winning amounts to across different games', with 'a range of properties (scoring more goals than the opponent, checkmating the opponent's king, etc.) that constitute winning for different games'. Likewise for truth: the properties that 'constitute truth' may be very different in 'different domains of discourse' – sometimes correspondence, sometimes superassertibility, sometimes coherence. Kevin Scharp (2013; 2021) does not think that there can be any concept of truth that satisfies all of the 'platitudes' normally associated with truth, opting instead for a pluralist project of conceptual engineering in which we try to craft a set of distinct alethic concepts that can replace truth.

In the remainder of this section I will give my own view on the different meanings and functions of 'truth' in different domains, drawing freely from the works of the authors mentioned above.[13] Table 4.1 gives a quick overview of the upcoming discussion. Going beyond identifying whole domains of discourse, my analysis will dig down into the level of particular

[12] See Lynch (2009); Wright (1992; 1999); and overviews in Pedersen and Wright (2018), Pedersen and Lynch (2018), and Glanzberg (2018a, sec. 4.4).
[13] I will also be building on my own previous work on 'truth and its multiple meanings' (Chang 2012a, pp. 240–3), which I am modifying significantly here.

types of activity in which each notion of truth is put to work. What I am offering is not a disjunctive taxonomy but several overlapping categories.

Truth-by-Assertion and Deflationism

The simplest activity relevant to the notion of truth is that of assertion-and-denial: to assert the proposition *P* is to say that *P* is true, and to deny *P* is to say that *P* is false. What 'truth' means in this activity is that a proposition is being asserted, regardless of whether one has good reasons for asserting it. Let's call it **truth-by-assertion**. These thoughts lead to the deflationary theory of truth. Since any statement is liable to being asserted, it would seem that the deflationary theory applies everywhere and it is the most basic idea of truth underlying any and all others. But it would be unwise to presume that all other notions of truth can and should be eliminated or reduced to the deflationary one.[14]

What exactly is this game of assertion, and when do we actually play it? Assertion *in the wild* is not the dry thing that it may seem in a logic class. There are other, more significant functions of saying something is true than a simple repetition. If I follow someone's utterance (including my own) by 'It's true', that is an act of emphasis, more effective and forceful than repeating the statement in full. If I say 'That's true . . .' in that intonation typically followed by '. . . but', that is an act of approving the bare content of what someone said while disputing some connotations or implications of it. Whether it is played bluntly or subtly, assertion is a game of personal *commitment*. When I assert something, I lend it whatever credibility I have, imply that I will act in accordance with it, and open myself up to others' judgement. That is to say, there are different illocutionary functions served by truth-by-assertion, while the narrow meaning of 'is true' is a simple affirmation of the proposition in question. So even the context of assertion may not be completely suited for the deflationary theory of truth.

Truth-by-Honesty and Truth-Telling

Another function of the word 'true' or 'truth' is to indicate the speaker's honesty or sincerity. The 'truth' involved here is a different kind of thing from the more objectivist conceptions of truth, as indicated by Price (1998; 2003) in his critique of the conflation between norms of sincerity and norms of truth.

---

[14] I also think it is too hasty to rule out non-assertoric functions of truth that may be served by unarticulated content that is only implicitly assumed.

The relevant activity for this notion of truth, which I will call **truth-by-honesty**, is that of truth-telling, or giving witness. Truth-by-honesty could be considered a kind of correspondence, between what I assert and what is in my mind, a match between what I say and what I perceive or think. But 'match' is not always quite right here, because the exact mental process going on in truth-telling is often not such a straightforward correspondence. In any case, truth-telling is operationally meaningful. 'I am telling you the truth when I say I saw a snow leopard on the hill.' This meaning is cogent whether or not my thoughts are true in some other sense ('A snow leopard in London? – surely not'). Truth-telling is not mere assertion, and a lie is also an assertion. Without the activity of truth-telling no epistemic community would be able to sustain itself, and no collective pursuit of knowledge would be possible. Truth-telling is not only a matter of personal character and disposition, but also an important part of the foundation of knowledge in any empirical system of practice.

Truth-by-Decree: Axioms and Conventions

Some statements are true because we *decide* that they shall be true. But such truth is only valid within our commitment to uphold it, and only interesting when we go on to *do* something on the basis of that commitment. I will call this **truth-by-decree**, the archetypes of which are definitions and tautologies. *Of course* it is true that the standard meter is one meter long (if it still *is* the standard of length, which it actually isn't any more in real metrology). And one cannot doubt the truth of the statement 'All bachelors are unmarried' since it is a tautology. These are truths that we construct, judge and maintain by making, using and enforcing specific meanings. They are also rendered true by presumption when we *take* them as given and engage in activities on their basis.

In mathematics we build systems of inference starting from axioms that are true by decree. Scientific theories can also be constructed and applied in an axiomatic manner. For example, when Einstein in 1905 postulated that the speed of light was the same regardless of the motion of either the observer or the source, this was a new axiom, which by no means followed from the previous definitions of 'light' or 'speed'. Another kind of truth by decree is conventions, both in the colloquial sense and in Poincaré's sense. Colloquially, a convention is something arbitrary that is socially agreed upon, like driving on one side of the road or the other, or when new technical terms or measurement units are defined. According to Poincaré, scientists decide to 'elevate' the best-confirmed empirical laws to the status of conventions, to put them beyond questioning. But even in that scenario truth-by-decree does not in itself amount to empirical truth.

Truth-by-Operational-Coherence

Now I come back to the kind of truth that I am most concerned with, which I named truth-by-operational-coherence in the last section.[15] (I will be brief here, since this concept has already been introduced, and will be elaborated further in the rest of the chapter.) This is the notion of truth at the heart of empiricism, embodying the idea that empirical truth is determined by what mind-uncontrolled realities are like, not by what we believe, say, do or wish. None of the other notions of truth discussed above captures this essential empiricist requirement. In connection with Price's theory (2003), it may seem that truth-by-operational-coherence only answers to the norm of personal warranted assertibility; however, to the extent that epistemic activities are conducted by communities of people, the Pricean norm of truth, which demands mutual agreement, is also in play.

Truth-by-Comparison

In Section 4.1 I drew a distinction between primary and secondary truth. Truth-by-decree and truth-by-operational-coherence are matters of primary truth, and truth-by-honesty could also be considered a matter of primary truth. Secondary truth is **truth-by-comparison**, taking 'comparison' in a broad sense. The grounding of secondary truth in primary truth can happen in various ways. Earlier I mentioned deduction and enumerative induction. If we want to be most liberal about secondary truth, we might allow it to be established by mere lack of contradiction with primary truths, or by inference to the best explanation. Interestingly, the traditional correspondence and coherence theories of truth both belong cogently in the realm of truth-by-comparison.

The Complex Functioning of the Notion of Truth in Real Practices

I hope that the distinctions made so far will be helpful in clarifying our thinking about truth in various domains of life. In typical situations in everyday life or scientific practice, various notions of truth are simultaneously in play, often in conjunction with each other. If we try to think about the whole conglomerate situation with just one undifferentiated notion of 'truth', we are bound to encounter difficulties.

---

[15] It is closely related to what I have called 'truth$_5$' previously (Chang 2012a, p. 242), but that notion also encompasses truth-by-comparison.

When different notions of truth are applied simultaneously to a given statement, the verdict of truth may vary. It often happens that a statement that is unwaveringly asserted or upheld honestly cannot support coherent activities. And it may be the case that Donald Trump often tells truth-by-honesty when he expresses his opinions (which is something that his supporters value), but the things he says really lack truth-by-operational-coherence in empirical domains. It can also happen that a proposition has a high degree of truth-by-operational-coherence, but is contradicted by a theoretical principle that is a truth-by-decree. In empirical science we routinely examine whether a postulate laid down as a truth-by-decree also has truth-by-operational-coherence; ultimately that is the process of empirical confirmation of theories. It is a wonderful scientific moment when truth-by-decree and truth-by-operational-coherence line up with each other, but that is by no means to be taken for granted. Poincaré's conventionalism plays it safer, by recommending that we ascertain the truth-by-operational-coherence of a statement to a high degree first, and *then* assign truth-by-decree to it.

It is also important to note that one and the same proposition may work both as a primary truth and as a secondary truth. Take the proposition that the atomic weight of chlorine is 35.5. This may be a primary truth determined by how well certain activities relying on this statement work out in chemistry and physics. For example, on this basis we can determine the amount of chlorine that will enter into chemical combination with certain amounts of other substances, and that works out well. But we may also establish the atomic weight of chlorine by measurement. In that case it is established as a secondary truth, grounded in the truth of certain physical assumptions underlying the operation of the measuring instrument. For another example, take the proposition that the surface area of a sphere is proportional to the square of its radius. This is a secondary truth in Euclidean geometry, provable from the axioms. But we can also take it as an empirical hypothesis, in which case it is established as a primary truth if we can rely on it for a set of coherent activities, ranging from figuring out the amounts of paint needed to paint balls of different sizes, to the Kantian deduction of the inverse square law of gravitation (Kant [1786] 2004, pp. 57–9; 518–21 in the original). Generally speaking, knowing how to manage different senses of truth is part of the art of inquiry.

4.3   Empirical Truth and Operational Coherence

Turning now to truth-by-operational-coherence, in this section I will focus on explaining and justifying the basic idea that truth consists in the facilitation of coherent activities, which will seem far-fetched to many people. Again I emphasize that my proposal of the notion of truth-by-operational-coherence is a *semantic* move concerning what primary truth should mean in empirical domains. Being relied on in coherent activities is *constitutive* of the truth-by-operational-coherence of a proposition, rather than being an effect or evidence of truth. I will also try to say more about what exactly is meant by the 'reliance' that coherent activities have on true propositions. The key point is that reliance is a practical kind of necessity: is the use of the proposition in question needed for the coherence of the activity in question? I also address the problem of coherent activities that rely on propositions that are not even intended as true, and argue that such propositions should be regarded to be true in relevant respects, or to an appropriate extent. Unlike the idea of correspondence to an inaccessible kind of ultimate reality, truth-by-operational-coherence is a concept that is in operation in all sorts of practices, including the actual procedures for theory-testing.

Coherence as Constitutive of Truth

Having set out various notions of truth and their uses in Section 4.2, I can now give a more focused discussion of truth-by-operational-coherence. Recall the brief definition of the notion that I gave in Section 4.1: a proposition is true to the extent that there are operationally coherent activities that can be performed by relying on it. I believe that this notion is the chief sense of primary truth operative in empirical domains, and so it provides crucial foundation for any empiricist epistemology or philosophy of science.

The first issue to consider more carefully is the semantic nature of my pragmatist proposal, to take operational coherence as *constitutive* of primary truth in empirical domains. Against this view you might want to maintain a more traditional stance, according to which truth is whatever it is (e.g., correspondence to ultimate Reality), and things like operational coherence are only relevant to the epistemic questions of how we know and learn truths. This approach amounts to separating out the metaphysics and the epistemology of truth; the epistemological side can be rightfully empiricist, while the metaphysical side remains with correspondence realism (as defined in Section 2.2). The common intuition is that an empirical statement possesses mind-independent truth, and that our epistemic activities such as hypothesis-testing are designed to *find out* this pre-existing truth. That is the intuition that I am trying to steer us away from.

We must not overplay the distinction between truth itself and the evidence for it, when it comes to truth-by-operational-coherence. I, too, want to preserve a distinction between truth and our personal knowledge of it, but taking truth to be separate from actual epistemic activities is not the way to do it. I think that is what James had in mind when he said ([1907] 1975, p. 37): 'The reasons why we call things true is the reason why they are true.' What it means for a statement to be true-by-operational-coherence is the same *sort* of thing as how we tell that it is true-by-operational-coherence. I propose to make the distinction between truth and knowledge as follows, in parallel with how I dealt with reality vs. our knowledge of it in Chapter 3:

- A proposition is true to the extent that there are operationally coherent activities that can be performed by relying on it.
- *I know* that a proposition is true to the extent that *I (personally) actually know how to carry out some* operationally coherent activities by relying on it.

So we can retain the role of operational coherence in the constitution of truth, while removing matters of personal circumstance from it. I fully expect that there are coherent activities that I have not yet learned to perform, or even conceived.

Truth-by-operational-coherence is the same thing as empirical confirmation, if we take 'confirmation' in a broad sense. Let's get back to some basic intuitions here. What should confirmation mean, other than surviving 'the test of experience'? And why should the test of experience be confined only to *explicit* testing, let alone testing that follows a particular formal scheme (e.g., the hypothetico–deductive model, Bayesian probability-updating, or null-hypothesis testing)? Explicit theory-testing might be the nearest activity-version of the idea of a statement corresponding to the world, but that is not a compelling reason to privilege it. If a theory can facilitate coherent activities, and not as a result of any strange accident or coincidence as far as we can see, then we can and should take (the relevant statements in) this theory to be 'true' – in the same down-to-earth sense as we say that it is true that rabbits have whiskers and live in underground burrows.

Theory-testing is not a distinct kind of activity from theory-use, but a subset of theory-use. Every attempt to use a theory may in principle function as an empirical test of it, and the truth or falsity of a proposition often emerges in activities that are not explicitly conceived as tests. 'The distance from Madrid to Barcelona is 15 kilometres' is untrue, and would be exposed as such, in that most activities we might try to carry out on its basis (such as walking from one place to the other in an afternoon, or hitting one from the other with artillery) would be operationally incoherent. Scientists in various scientific fields ranging from epidemiology to cosmology are very familiar with the situation in which no explicit tests can be devised, and hypotheses just have to be evaluated by how well they account for observed phenomena 'in the wild'. (The process of validating new vaccines goes slowly because we do not have human guinea pigs to whom we can deliberately give the pathogen after giving them the vaccine, but that is just a constraint we have to live with, unless we want to change the fundamental ethical basis of social life.) And even the classic hypothetico–deductive method of theory confirmation involves *using* the hypothesis to derive predictions, which we then evaluate by comparison with observations. Theory-confirmation is a measure of how coherently this activity works out, of making empirically accurate predictions relying on the content of the theory under test.

Any positive test-outcomes partially *constitute* the theory's truth;[16] it would be a category-mistake to say that positive test-outcomes are *caused* by the theory's truth. James warned against the reification of truth as with other 'words in *th*' (wealth, health, strength and truth), quoting an epigram from Lessing: 'How come it's exactly the richest people in the world who have the most money?'[17] There is, of course, nothing to explain here, because having lots of money is exactly what being wealthy *means* (James [1907] 1975, pp. 105–6). So it is, too, with truth-as-operational-coherence. Here again I am rejecting the relation between empirical success and truth envisaged in standard scientific realism, according to which truth explains success. It may be instructive to contrast my view with Kitcher's, as he tries to preserve an explanatory role for correspondence truth. Although he conceives the realist success-to-truth inference in a pragmatist or activity-based way, Kitcher still preserves the inference from '*S* plays a crucial role in a systematic practice of fine-grained prediction and intervention' to '*S* is approximately true' (Kitcher 2012, p. 112). In more recent work, he takes this success–truth link as a 'plausible empirical conjecture'; if the conjecture holds, it forms the basis for treating truth as the explanation for success (Kitcher forthcoming, ch. 3). In contrast, I do not take the success–truth relation as a matter of inference or explanation; instead I take success (or rather, operational coherence) as the core of the very meaning of 'approximately true', or rather, 'true (to a degree)'.

In James's evocative expression, 'truth *happens* to an idea. It *becomes* true, is *made* true by events.'[18] This is a pointer to what happens as

---

[16] A finite number of positive test-outcomes will not add up to the complete truth of a general theory; this is like the relation between instances of empirical success and the full empirical adequacy of a theory, in terms of van Fraassen's epistemology.

[17] 'Wie kommt es, Vetter Fritzen / Dass grad' die Reichsten in der Welt / Das meiste Geld besitzen?'

[18] James [1907] (1975, p. 97) emphases original; quoted in Kitcher (2012, p. xxiii).

the idea goes on to facilitate further coherent activities. Truth-by-operational-coherence consists in the significant roles that propositions play in our various activities, and more such roles emerge as we continue to learn to do more things. In an imperfect yet suggestive metaphor: think of a proposition attaining truth like a child growing up. Even though we would say that a particular child has the potential to grow up to be a certain kind of adult, the adult is not fully *there* in the child. So with truth. We can allow that there are coherent activities as yet unknown to us, but when we have to make actual concrete judgements of truth we have nothing else to go on except the truth that is already known in activities that we can actually perform.

The Meaning of Reliance

Return to the definition of truth-by-operational-coherence: a proposition is true to the extent that there are operationally coherent activities that can be performed by relying on it. What exactly does it mean for an activity to 'rely on' a proposition? In evaluating the truth of a proposition, we should ask: can the coherence of the activity in question be maintained, if we do not let the proposition in question inform what we do? The sense of 'reliance' incorporates a practical and empirical need,[19] a sense that the activity in question cannot, as a matter of fact, be performed without making use of the proposition in question. For example, most problem-solving in classical electrodynamics cannot be done without relying on Maxwell's equations. The notion of 'reliance' is meant to capture the sense in which the employment of a proposition is productive and meaningful. For what I have in mind by 'employment' here, it is not sufficient that a proposition is asserted in the course of an activity. So we can perfectly well use Maxwell's equations while denying that the aether exists, even though many pre-Einsteinian physicists did frequently assert that the aether existed when they were using Maxwell's equations. The activity of solving Maxwell's equations for various problems in electrodynamics does not rely on the proposition that the aether exists. Insisting on actual employment also guards against the possibility that the proposition in question might be involved in the activity in a superfluous way reminiscent of the tacking paradox or the Gettier problem, in which case we would not want to make an attribution of truth.

---

[19] In a previous formulation I expressed the idea as follows: 'a statement is true in a given circumstance if (belief in) it is *needed* in a coherent activity' (Chang 2017b, p. 113). What I meant to express by the term 'need' there was not logical or metaphysical necessity, but a pragmatic necessity – what we actually need, which can only be learned empirically. Now I propose to avoid invoking a notion of necessity, since it is liable to cause confusion.

Reliance was the issue, for example, at the core of my debate with Psillos regarding the successes of the caloric theory (Chang 2003), mentioned briefly in Section 2.4. I think I have shown to most people's satisfaction that some very coherent activities of thermal physicists around 1800 did rely on key tenets of the caloric theory. For example, Pierre-Simon Laplace's derivation of the ideal gas law relied on assumptions about the material reality of caloric, about its corpuscular constitution, and the basic nature of the forces operating between particles of caloric. Psillos had stated that no substantive assumptions about the nature of caloric were relied upon in the successes of the caloric theorists, and I disagreed. Which one of us was right about the historical facts is not the main issue for my present purposes. I think we both agreed that this kind of reliance, which we can judge by a concrete examination of the practices of scientists, is what underpins the truth of the propositions in question. To the problem of suspected superfluous propositions there is no magic solution, only the hard work of empirical investigations. This sort of checking may not live up to some overblown image of a philosophical test, but it is how we get on in science, and in the rest of life, too.

It may be objected that reliance is too loose a notion, especially as it would seem to incorporate psychological factors. Doesn't it sometimes happen that people rely on certain propositions in some psychological way that should not be considered to have any bearing on truth? What if our mental make-up is such that belief in a certain fantastical proposition is psychologically necessary for us to carry out some activities? For example, what if I can only swim by believing that I am a dolphin? Is it then true that I am a dolphin, because my coherent activity of swimming relies on that belief? On the face of it, this seems like a straightforward and devastating objection to my notion of truth-by-operational-coherence, or any pragmatist notion of truth for that matter. Actually the problem arises from positing an inherent connection between truth and *belief*. Thinking in terms of reliance on content instead of belief allows us to avoid unnecessary tangles with psychologistic considerations. Psychological reliance on a belief is a different issue from the employment of a proposition in an activity. Imagine a nineteenth-century scientist who can only get enough motivation to do his work in optics because he believes in the reality of the noble luminiferous aether, although he does not rely on any particular assumptions about the nature of the aether in his reasoning.[20]

---

[20] I thank Mike Martin for raising a version of this worry at the Aristotelian Society in January 2017. His example was about delusional self-confidence that allows scientists to tackle very challenging problems (successfully).

Coherent Use without Truth?

Now, there may be times when people put to good use some propositions which they expressly do not regard as true. This is a common situation in scientific research, for example when admittedly false models are used in various coherent epistemic activities including prediction and explanation (see Suárez 2009; Toon 2012; Rowbottom 2019). Doesn't this make trouble for my notion of truth-by-operational-coherence, according to which propositions relied upon in coherent activities are true? It would appear that my account faces a threatening conceptual tangle in the idea of a coherent use of an untrue proposition, while more standard accounts of truth can freely allow that useful and even correct consequences can sometimes follow from falsehoods. For several different reasons, this is not a problem.

A relatively simple type of case is where we may declare a theory or model false on the whole, but actually trust or affirm the aspects of it that are being actively relied upon in coherent usage. And then there are cases in which we regard the theory or model in question not to be strictly true but 'true enough', and also freely allow that its uses are only coherent enough, so there is no glaring mismatch between coherence and truth. Or we may make a working commitment to hypotheses whose probability would have to be assessed as not comfortably close to 100 per cent. Hypotheses are very often *used* in the absence of conclusive confirmation, as we should know from the existence of the phrase 'working hypothesis'. In fact, life is very much a game of relying on less-than-certain hypotheses, which are open to doubting and testing. For example, we may count on the victory of a leading candidate in an election, assume that a currently dominant scientific theory will not be rejected for some time, or even plan the next year on the basis that we will not be involved in car crashes or contract serious illnesses. Such working hypotheses are relied upon provisionally because we do not have a better basis for our activities. But if these activities turn out to be coherent, then the working hypothesis does earn a measure of truth-by-operational-coherence. If not, our commitment is withdrawn. So the mismatch between truth and coherent use would be quite temporary.

It may also happen that the proposition in question is true in one sense of truth, but not true in another sense of truth (recall the various senses of 'truth' discussed in Section 4.2). For instance, it can easily happen that a theory or model lacks truth-by-comparison because it conflicts with another that is more firmly established as true, but it possesses truth-by-operational-coherence through its usage. For example, as mentioned in Section 3.4, the theory of atomic and molecular orbitals supports many extremely coherent activities in chemistry, but is declared to be false by later versions of quantum mechanics that treat electrons as identical (indistinguishable) particles. In such cases, it would be wrong to say that the successfully used model is *simply* false. It is true-by-operational-coherence at least in a limited domain.

Sometimes we take a theory or model as only apparently propositional. We may even subscribe to a generalized instrumentalism, which treats theories or models as tools of reasoning rather than sets of propositions. In that case the question of truth does not arise.[21] That is a tenable position, though I prefer a realist–pluralist outlook that attributes some truth-by-operational-coherence to each reliably successful model, as I will explain further in Chapter 5. Or we may take a fictionalist attitude. Think back to the Newtonian physicists with their supreme confidence in the truth of Newton's laws of motion and gravitation. Now we look back and say that Newton's laws are not entirely true, but we can make fictionalist sense of Newtonian practice, or in a surrealist way, as Tim Lyons (2003) would have it: the world, in many domains, behaves *as if* Newton's laws were true. So here is a kind of pessimistic induction: who is to say that our current most-trusted theories will not turn out to be only fictionally true? But I would prefer to say that they *are* true within the recognized limits. This amounts to an optimist–realist answer to the pessimistic induction: within the successful activities that we carry out, there is truth that should not be negated by a later fictionalist account of the situation.

4.4   Truth as a Quality

One key feature of the concept of truth-by-operational-coherence is that it is not an absolutist notion. In my opinion this is both important and beneficial, so some elaboration is called for. My definition of truth-by-operational-coherence states that a proposition is true 'to the extent' that it facilitates coherent activities. 'To the extent' there was a deliberately vague designation for the sake of brevity, and now I must say more about what it means. There are two main aspects, which I will address in turn. First, the truth-as-operational-coherence of a proposition comes attached with a specific and finite scope, given by the set of coherent activities that relies on it. Secondly, the extent of truth depends on the number and variety of the coherent activities, the degree to which each of those activities relies on the proposition in question, and the degree of coherence of each activity. It

---

[21] This is different from constructive empiricism (van Fraassen 1980), which regards statements concerning unobservables as truth-apt while recommending against trying to find out their truth-values.

will be futile to try to summarize all of those considerations into a single measure. So truth-by-operational-coherence is not a matter of binary true/false, not even a simple matter of degree, but a multi-faceted and contextual *quality*.

The Finite Scope of Truth

In unpacking the sense in which truth-by-operational-coherence obtains 'to the extent' that it facilitates operationally coherent activities, the first thing to note is that truth has a finite scope. This is a very useful insight when we are trying to make sense of the history of science (or of any other evolving field of knowledge). We do often say that certain theories are only true within a certain domain. The special theory of relativity is true only where there is no gravitation. Ordinary quantum mechanics is true enough in the world of atoms and molecules, but it is not designed to handle the interaction between matter and light, or the collision of elementary particles at high energies. As noted in Section 4.1, there is a temptation to say that all theories are false except the most cutting-edge one that we now believe, but that would be to walk right into a trap, since we have to grant the possibility or even likelihood that our current best theories will later be shown to be false, too. This makes 'truth' a basically unusable designation. The only sensible way to shield ourselves from the pessimistic induction from the history of science is to grant that all good theories are true, and remain true, within their own scope. And with my notion of truth-by-operational-coherence, such liberal granting of truth is not only possible but very natural. (This has a clear implication of pluralism, which will be discussed further in Sections 4.5 and 5.4.)

Now, the usual tendency among scientific realists is to assume that later and better theories will be true in larger domains, and that this is a cumulative process, so that the domain of a previous theory is completely included in the domain of a later and better theory (recall Einstein and Infeld's image of the ever-expanding vista as one climbs a mountain, discussed in Section 2.5). If that is the correct picture of the history, then the limited scope of the truth of a current theory becomes bearable in the optimistic hope that we are approaching, step by step, 'a final theory, one that would be of unlimited validity and entirely satisfying in its completeness and consistency' (Weinberg 1992, p. 3). But there are good reasons to be sceptical about this picture (even if we set aside the problem of incommensurability, which I will discuss in Section 4.5).

We tend to have an overblown sense of how well later theories account for the successes of earlier theories in science. For example, so often we hear that quantum mechanics is true everywhere, while classical mechanics only approximates it well enough in macroscopic domains. But we actually have no *direct* empirical evidence that quantum mechanics is true in the standard macroscopic domains where classical mechanics does its best. It is considered a stunning achievement to demonstrate unequivocally quantum-mechanical effects even in bodies just large enough to be above the threshold of what we can see with a microscope, for example when two Bose–Einstein condensates come together and make an interference pattern. Even in such amazing cases, what we have is only a qualitative demonstration of a quantum effect, not the confirmation of a precise prediction made from a quantum-mechanical calculation.[22] When it comes to typically macroscopic situations concerning, say, the trajectory of a rocket ship, or a spinning top, it would be a mind-boggling technical problem to figure out how to adapt the Schrödinger equation to such objects. Now, if someone figured out how to do that, can we presume that the quantum-mechanical calculation would generate empirically correct results, and more correct than the result of the classical-mechanical calculation? When I raise this question, the usual response I get is simple incredulity: how *can* you question such a thing? It is a common article of faith that quantum predictions will be truer than classical predictions *everywhere* because quantum mechanics is a generally superior theory to classical mechanics. But how is the assumption of such a uniform superiority justified? (If you are a better person than I am, are you better at *everything* than I am?) As for the elaborate accounts given on how quantum descriptions approach classical ones in the macroscopic realm, ranging from Ehrenfest's theorem to decoherence, they only demonstrate certain sorts of convergence between the two theories, but cannot in themselves say which theory has the empirical edge where they differ. In my opinion, it is better to recognize that classical mechanics is still true where it always has been, and that its truth does not consist in a successful reduction to quantum mechanics. The same point could be made even more strikingly if we considered the relation between classical mechanics and quantum field theory or the standard model of elementary particle physics.

Let us try to discern the scope of truth-by-operational-coherence where it actually obtains, without getting distracted by the 'dreams of a final theory'. For another example, take what was perhaps the single most important proposition in the history of organic structural chemistry in the nineteenth century: 'Carbon has valency 4', meaning that it is capable of bonding with

---

[22] See Tomczyk (2022) for an insightful historical and philosophical discussion of Bose–Einstein condensates.

four other chemical units (atoms or radicals) at the same time. At least to begin with, this proposition was a truth-by-operational-coherence: there was no way to compare it directly with observations, and the theories of the internal structure of atoms from which it could have been deduced were still many decades away. But the highly productive activity of working out the molecular structures of countless organic substances relied crucially on this tetravalency of carbon. It was also relied upon in the understanding and planning of countless chemical reactions, especially 'substitutions'. For example, a body of methane gas could be made to absorb a volume of chlorine gas and emit an equal volume of hydrogen gas, turning the methane into chloromethane; such a substitution could be made four successive times, in the end yielding what we call carbon tetrachloride. All this makes perfect sense when we use the tetravalency of carbon to understand methane as CH_4, chloromethane as CH_3Cl, and carbon tetrachloride as CCl_4 (see Figure 2.1 in Section 2.5 again). Such highly coherent activities in analytic and synthetic chemistry were what constituted the truth of 'Carbon has valency 4', and those activities still maintain their coherence to this day. But this truth is a limited one. We know, for example, that the structure of carbon monoxide remained a mystery for a long time. Even carbon dioxide (CO_2) was not trivial to understand, but it could be accommodated by saying that the carbon atom in it formed a *double* bond with each of the two atoms of oxygen (which has valency 2), thereby using up all of its 4 bonding-potentials, as indicated by the graphic formula O=C=O. But it was not clear at all how carbon monoxide (CO) could be understood until quantum chemistry arrived with its sophisticated and entirely different account of chemical bonds. And in the new theoretical regime the old concept of 'valency' as such had no place (and the sticks that represent the bonds in the old ball-and-stick models have no straightforward quantum-mechanical equivalents). However, the nineteenth-century valency values are still true, to the extent that they continue to enable a myriad of highly coherent analytic and synthetic activities of chemistry.

The recognition that any given truth has a limited scope is also crucial for dealing with what I will call the 'objection from effective false beliefs' (essentially the same as what I called, less soberly, the 'witchcraft objection' in Section 4.1). The objection is quite simple: we know that false beliefs can sometimes lead to success, so we should not equate truth with success (or rather, with what leads to success). The same sort of objection can easily be directed to my notion of truth: in a whole range of cases, including the reality of witches and various attributes of God, it would seem that people can engage in *some* operationally coherent activities by relying on false beliefs. For example, on the basis of the assumption that the earth is flat, one can carry out a lot of quite coherent activities; surely that doesn't mean it is true that the earth is flat? But we need to be careful about this kind of case. Coherent flat-earth activities rely on the assumption of local, not global flatness, which is true enough; there is no harm in allowing truth to that (see Teller 2021, p. S5023, footnote 20). Now, global flatness can also be 'held true come what may', but this becomes a classic degenerating research programme in the Lakatosian sense. After all sorts of excuses are made and risky tests are declined, the coherence of global flatness that remains is in a pitifully tiny domain excluding images from satellites, the global positioning system, the existence of Antarctica, etc. It is very instructive to read Lee McIntyre's account (2021) of his efforts to engage flat-earth advocates in an empirical discourse.

When we say that mere convenience should not be mistaken for truth, that is normally because 'the truth will out' (recall James's image of experience 'boiling over', cited in Section 1.6). That is to say, we should not assert that a statement is 'true' without qualification, if we expect that it might be shown not to be true in some other circumstances. When we say 'It may seem as if P were true in these circumstances, but P is actually not true', what else can we be meaningfully asserting, other than that P will fail to be true in some other circumstances? Those who invoke the objection from effective false belief tend to be overly fearful of easy and *undeserved* coherence, and it might be reassuring to be reminded of just how difficult it is to devise and maintain coherent activities in real life. They also tend to be overly dogmatic in denying truth to propositions at the basis of various coherent activities. I have already argued that it is not wise to deny truth altogether to the best scientific theories from the past. I see a disturbing pattern of thinking: one makes up one's mind, a priori, about the falsity of some doctrine; when there are coherent activities relying upon it, instead of allowing that the doctrine might be true in those circumstances, one concludes that facilitating operational coherence cannot be what truth is about, retreating to the empty notion of truth by correspondence to ultimate Reality instead.

Truth as a Quality

By saying that a proposition is true *to the extent* that there are coherent activities relying on it, I also mean that its truth is a matter of degrees, depending on how well it plays this role. Actually it is not quite right to say 'matter of degrees', because 'degree' has a connotation of a numerical quantity, and 'how well' here is not easily quantifiable, at least not into a single number. There are many factors that feed into truth-as-operational-coherence. What is the number and variety of the coherent activities that rely on the proposition in question? How strong is the reliance that each of those activities has on the proposition? How coherent is each activity? And if we were to try to get an overall measure of the degree of truth, we would have to also consider how valuable each of those activities is, so they can be given proper weights, and there would also be a need to balance out coherent and incoherent activities attempted within the same domain. Truth-by-operational-coherence would at best be a multi-dimensional quantity, or an index composed of several quantities, but even such a move could be worked out convincingly only if each dimension is properly quantifiable.

Instead of trying to construct an elaborate and unworkable quantitative measure, I think it would be better to accept that truth-by-operational-coherence is a *quality*, of which we can discern a sense of 'more' or 'less' only in an imprecise way. Truth-by-operational-coherence should be seen as yet another case in the class of qualities that allow comparisons but not in unequivocal and fully quantitative ways. There are a number of such qualities. With such qualities, sometimes there are clear more-and-less judgements, as in the hardness of diamond versus talc. Some cases are much more equivocal: glass is harder than iron according to the what-can-scratch-what criterion of the Mohs conception of hardness, but it is clearly less hard than iron in the sense of being more fragile. Some other qualities only allow comparisons in extreme cases or in very specific contexts. We can confidently say Mother Theresa had a higher degree of benevolence than Adolf Hitler, but it will be difficult to be definitive about John F. Kennedy versus Franklin D. Roosevelt. Truth-by-operational-coherence is akin to these latter cases. And perhaps it is only appropriate that truth comes out looking like other major virtues in life, such as beauty, justice and benevolence.

It may be useful here to consider some other attempts that have been made to acknowledge that empirical truth is not a black-and-white binary judgement. The idea of approximate truth is often invoked, especially by scientific realists, but hardly ever precisely defined. In a similar direction serious work has been done on the quantification of the truth-likeness (verisimilitude) of theories, starting with Popper's attempts and evolving into versions like the one given by Niiniluoto (1999, sec. 3.5), ultimately based on the idea of counting up the number of true and false consequences of a theory. But such quantification is workable only because it rests on the assumption that the truth and falsity of the consequences of a theory are themselves unequivocally defined. Another popular move in recent epistemology has been a resort to probability. But it is difficult to reduce truth-by-operational-coherence to probability, because it is not a matter of well-defined frequency or degree-of-belief. And even if we could somehow squash the notion of truth-by-operational-coherence into one dimension, it does not sit well with the kind of quantification as probability, plotted on a scale of 0 to 1.

Admitting truth to be a quality raises a very significant logical issue, to which I cannot do anything like full justice here.[23] With truth as a quality, what happens to all the rules of logic based on the strict mutual exclusion of true/false? As indicated in Section 1.6, I think we should take this as a serious warning that bivalent logic is not the ultimate guide to real life. It is an interesting quandary to think what 'false' should mean when 'true' is a quality. My very amateurish view, for the purpose of the philosophy of science, is that 'true' and 'not true' should still be taken to be mutually exclusive just to preserve the meaning of 'not', though that isn't all that useful when 'true' itself is a qualitative notion. And 'false' could be given a more interesting meaning than 'not true'. I think there is much use for a notion of falsity which can be given to a proposition that would, if relied upon, actively destroy the coherence of existing activities. Under some circumstances we can work conveniently treating truth and falsity as if they were simply mutually exclusive categories, but we should not mistake that as the ultimate and universal rule of thinking.

As indicated by the logical questions just mentioned, I realize that the idea of applying the term 'truth' to what is less than perfect and universal will grate against many philosophers' intuitions. To soothe the irritation, it will be helpful to consider more carefully some of the numerous instances in which we freely and confidently assign truth to statements that we know to be quite likely to be imperfect and limited. In fact, I would argue that most cases of truth-assignment in any real practices outside classical formal logic are of this type. For the purpose of having a philosophy that is equipped to make sense of real life, it is actually a good thing that our concept of truth does not designate something perfect. 'True enough' may be the operative criterion in many practical and scientific decisions. As James once put it: 'Any idea upon which we can ride, so to speak ... is true for just so much, true in so far forth' (James [1907] 1975, p. 34).[24]

4.5   Plurality and Incommensurability

If truth-as-operational-coherence is qualitative in the sense explained in Section 4.4, it becomes natural that there may be multiple true

---

[23] I thank Dorothy Edgington for making me realize that I could not avoid this issue.

[24] In the remainder of this passage James states that an idea is 'true instrumentally'; this is a fundamentally different attitude from what usually goes by 'instrumentalism' these days, which amounts to denying truth-value to propositions regarded as instruments.

theories[25] in a given domain. This plurality is a separate issue from the recognition that there are multiple *concepts* of truth. But doesn't the plurality of truth-as-operational-coherence lead to contradictions? I argue that this is not a worry at least in most cases, because successful competing theories generally employ sets of concepts that are sufficiently different from each other, so that propositions in the competing theories do not directly contradict each other. This is an expression of Kuhn's doctrine of semantic incommensurability. The recognition of incommensurability leads to epistemic pluralism, to go with the metaphysical pluralism expressed in Section 3.4: it makes sense that scientists tend to maintain a plurality of theories by conserving successful theories even after the emergence of alternative theories that are superior in some respects.

The Plurality of Truth-by-Operational-Coherence

I expect that my views expressed so far will have raised a worry about relativism in many readers' minds. I do not think that relativism, in the sense of recognizing the absence of absolutes, is to be feared so much, but at any rate the implications of my views on truth are more pluralist than relativist (Chang 2020b). My perspective on truth is pluralist at two levels. First of all, I think there are many valid and useful *concepts* of truth, as discussed in Section 4.2. At another level, under the concept of truth-by-operational-coherence, which is my main focus here, we see that multiple truths can hold concerning the same subject-area. It is this latter kind of plurality that I want to discuss further now.

The first point I want to stress is that the plurality of truth-by-operational-coherence does not necessarily imply contradiction, and that it very rarely does in practice. The plurality of theories is threatening only if we assume that each theory is complete and perfect. As discussed in Section 4.4, the scope of the actual use of theories is often narrower than imagined, and they do not overlap as much as we might fear (or hope). Take, for example, the business of wave–particle duality in quantum mechanics. Niels Bohr recognized the truth and necessity of both wave-descriptions and

---

[25] Instead of theories it might be better to think about 'accounts' as defined by Catherine Elgin (2017, p. 12): an account 'consists of contentions about a topic and the reasons adduced to support them, the ways they can be used to support other contentions, and higher-order commitments that specify why and how evidence supports them. It also contains normative and methodological commitments specifying the suitability of categories, the criteria of justificatory adequacy, and the ways to establish that the criteria have been met.'

particle-descriptions of quanta, and generalized this insight into his doctrine of complementarity (see Murdoch 1987, chs. 1–5). Even if we do not accept the deeply metaphysical aspects of Bohr's view, it must be agreed as a matter of fact that electrons and other quanta exhibit particle-like behaviour under certain experimental conditions (as in the photoelectric effect or the Compton–Simon experiment), and wave-like behaviour under certain other experimental conditions (as in the double-slit experiment). A vivid image of this wave–particle duality was given by Bruce R. Wheaton (1983, epigraph), quoting J. J. Thomson from 1925: 'It is like a struggle between a tiger and a shark, each is supreme in his own element, but helpless in that of the other.' Even though we are dealing with the same objects in one sense, we are looking at very separate sets of phenomena that are exhibited in very different experimental activities that we engage in. We could quite plausibly say that the wave-like description is true in one domain of phenomena, and the particle-like description in another, even though we can also discern the same entities that are involved in both sets of phenomena.

But wouldn't such breezy pluralism raise a difficulty with my own notion of truth? The definition of truth-by-operational-coherence requires that a true proposition should be relied upon in coherent activities. But if the same job may be done in two different ways, neither is *necessary*. So how can either be said to be *relied upon*? For example, if the same experiment can be interpreted in two wholly different ways, wouldn't that make all the statements in either of the interpretations unnecessary? This is where the subtle difference between reliance and a stronger notion of necessity comes in. I take reliance as a matter of *actual* need *within* a given system of practice: so we might say, for example, that having adopted the caloric theory of heat and the explanatory practices that come with it, I have to rely on the assumption of the self-repulsion of the caloric fluid in order to be able to explain the phenomena relating to the pressure of gases (for example that it increases with temperature). But if I do not work in the calorist system at all, then of course my explanation of pressure does not rely on any assumptions about caloric; if I am working with the kinetic theory of gases instead, then the same phenomena can be explained by relying on the assumption that gas molecules are bouncing around at random and that temperature is proportional to the molecules' kinetic energy. So both the assumptions of the caloric theory and the assumptions of the kinetic theory have some truth-by-operational-coherence. To take another example briefly: there are coherent activities that I can perform using the phlogiston theory (which Priestley and others did perform), and there are also coherent activities that you can perform using Lavoisier's oxygen theory. I am relying on the phlogiston theory, and you are relying on the oxygen theory; each theory is true to that extent.

Incommensurability Revisited

The type of situation just discussed, in which we have very different theories being used in the same domain, raises a different type of worry for my view on truth as well. Won't we end up with contradictions in such cases? The phlogiston and the oxygen theories do say things that apparently conflict with each other, for example whether water and metals are elements (the phlogiston theory says that water is an element and metals are not, and the oxygen theory says the opposite). Or consider the case of light again. Light is taken as a collection of rays that obeys certain rules of reflection and refraction (geometric optics), or as waves in the aether that can exhibit diffraction and interference (the original wave optics), or as patterns of propagation of intertwined electric and magnetic fields (modern Maxwellian electrodynamics), or as a bunch of photons each carrying a little bundle of energy (Einstein's quantum theory of light), or as photons that mediate electromagnetic forces between charged particles (Feynman's quantum electrodynamics), or excited states of a quantized field (quantum field theory). All of these conceptions are at least in principle applicable to every situation where light is present, and each conception embodies a set of truths. While some combinations of these conceptions are quite harmonious, some other combinations are not. So how can we avoid contradictions, if we grant truth freely and pluralistically as I am proposing?

I would argue that in most such cases in actual science there is still no contradiction. In order to unpack and justify that claim, I need to enter into a discussion of incommensurability. There is clearly a kind of dissonance, a lack of agreement, between saying that light is a continuous wave-like distribution of electric and magnetic fields, and saying that light is a collection of discrete photons each carrying a definite amount of energy. But no physicist would deny that discrete photons pass through the slits in the double-slit experiment, or that there is an electromagnetic wave hitting the metallic surface in the photoelectric effect. In situations like this, we have to admit that we have two parallel descriptions, which are related to each other (for example, the energy of a photon is proportional to its *frequency*, a notion rooted in the wave-description) yet not fully translatable to each other. This is a clear case of semantic incommensurability as discussed by Kuhn.

Kuhnian semantic incommensurability does not imply contradiction. This issue is complex and deserves further consideration (see Hoyningen-Huene and Sankey 2001 for a thorough discussion). To continue with the example of light: at the general level of description, the idea of photons as discrete packets of energy does seem to contradict the idea of an electromagnetic wave which carries energy in an entirely continuous manner. But when it comes to the level of specific statements, the contradiction often evaporates into a mist of untranslatability. For example, how do we render into wave-theoretic language the statement 'One of five photons, each of energy E, travelling along the x-axis, was absorbed by an atom at location L and thereby put the atom into an excited state'? We can try to depict a plane wave propagating in a definite direction, but in that case the wave theory has no terminology for expressing the discreteness of photons, or localizing the wave to the location L. The statement in question cannot be faithfully rendered into wave-theoretic terms, which means that we cannot even ask what the wave theory would say about that situation, to see if it would contradict what the particle theory says.

Therefore, if we take the early twentieth-century wave- and particle-descriptions of light and ask which is true, the only sensible answer is that they are each true, to an extent. Which one is truer would be difficult to say, as they do better in different situations (recall 'a tiger and a shark'). How should we choose between the two, then? That was the very controversial question of theory-choice (or paradigm-choice) that Kuhn's work raised. The pluralist answer is that we do not need to *choose* between the alternatives in a way that preserves only one and kills off the other. We keep both descriptions, and use one or the other or both, wherever each can facilitate coherent activities. By facilitating coherent activities each description acquires truth. The actual history of physics bears out this possibility; quantum physicists learned to live with wave–particle duality, even if they did not all follow Bohr on the general doctrine of complementarity. In earlier optics, too, the particle theory and the wave theory co-existed and each demonstrated some truth. Many scientists saw this situation as one of mere ignorance and indecision, and expected the debate to be resolved in favour of one theory or the other, but the subsequent history did not bear out that monist expectation. Think back to the inspiration that Kuhn took from gestalt psychology. Is the duck–rabbit really a duck, or a rabbit? Does the Necker cube drawing really depict the cube one way, or the other? Both interpretations are correct, depending on how we are looking at it. There may be contextual factors which make one more apt than the other (for example if the duck–rabbit figure is part of a depiction of a duck farm), but there is no sense in which one can be absolutely true and the other one false. Less metaphorically, when there are mutually incommensurable descriptions, the truth of each depends on how coherent its uses are, but *not* on how coherent the uses of other descriptions are.

4.6 Rehabilitating the Pragmatists

Having laid out my own ideas about the nature and functions of the concept(s) of truth, I now want to make some reflections on the so-called pragmatic theory of truth. It is important to get this aspect of pragmatism right, because truth is the issue on which classical pragmatism was most seriously misunderstood and attacked. The classical pragmatists' views on truth should not be caricatured as a notion that whatever pleases the believer is true. In particular, I argue that William James's controversial notion of truth can be rehabilitated through my notion of truth-by-operational-coherence. I will also try to dispel some common misunderstandings of the pragmatic theory of truth. To start with, it is important to note that pragmatism does not equate truth and utility. More subtly, according to my interpretation what the pragmatic theory offers is a shift in how we conceive the very meaning of truth, not merely a new set of criteria by which truth can be judged.

What Is the Pragmatist Theory of Truth?

In this section I want to make a defence and productive reinterpretation of the classical pragmatists' views on truth. The account of truth was a central part of pragmatism, especially as James presented it, and it was also where pragmatism drew the fiercest criticism. The classical pragmatist notions of truth are not far from my notion of truth-by-operational-coherence, which is no surprise since my thinking has been strongly inspired by the pragmatists. I cannot enter into an in-depth discussion of the classical pragmatists' views here, but given the importance of pragmatism in the general orientation of this book, it would be remiss of me not to comment on how I understand the chief pragmatists' views on truth. I also believe that looking back on the classical pragmatists through the lens of my own ideas can suggest a productive reinterpretation of their thoughts.

There are two main streams of thought on truth within the pragmatist tradition. One is due to Peirce, who led the common pragmatist aversion to the typical correspondence theory or any other 'transcendental' accounts of truth referring to inaccessible metaphysical realms, which made truth 'a useless word' (Peirce 1934, §5.553). But what exactly was the pragmatic 'upshot' (Peirce 1934, §5.4) of the concept of truth? Wanting to preserve a sense of objectivity in the idea of truth while not falling back into correspondence realism, Peirce famously opted for the following formulation: 'The opinion which is fated to be ultimately agreed to by all who investigate, is what we mean by the true, and the object represented in that opinion is the real' (Peirce 1986, W3.273). Similarly, in a later formulation: 'Truth is that concordance of an abstract statement with the ideal limit towards which endless investigation would tend to bring scientific belief' (Peirce 1934, §5.565). So, when all is said and done, at the end of all inquiry, a community of sincere and able inquirers will be left with a set of truths.

There is an obvious futility to the idea of 'the end of inquiry' for anyone wanting to stay true to the basic spirit of pragmatism. The ultimate end of inquiry that Peirce envisages here is quite as inaccessible as the metaphysical realist's 'external world' or 'Reality'. So Peirce's notion of truth is just as useless as the correspondence notion of truth, and detracts from Peirce's own spirit of pragmatism. I am inclined to agree with Cheryl Misak's argument (2007b) that the real core of Peirce's notion of truth was not the idea of convergence or the fate of knowledge in the long run, but the idea of 'indefeasible belief'. Indefeasibility means that the idea under consideration 'would not be improved upon; or would never lead to disappointment; or would forever meet the challenges of reasons, argument, and evidence' (Misak 2013, p. 36; see also Misak 2007b, p. 68). This is surely correct as a reading of Peirce's later works, but it still does not solve the problem that worries me. Indefeasibility, in the sense of being able to meet certain challenges 'forever', is not something we can evaluate in the present at all, any more than we can foresee the outcomes of inquiry at the ultimate end of the process.

Thus setting aside Peirce's view for the moment, I now come to the other main stream of the pragmatist theory of truth, which is the more controversial view usually attributed to James. This view requires a careful exposition and defence. James himself ([1907] 1975, p. 95) called it the 'Schiller–Dewey view of truth', but I will focus on James's presentation here, not only because it is the best-known version but because he (unlike Dewey) considered the elucidation of the truth concept such a central issue, viewing it as one of the two key components of pragmatism, the other being 'the pragmatic method' (ibid., p. 37). He gave significant attention to the concept of truth in his flagship text *Pragmatism* published in 1907, and followed that up two years later with *The Meaning of Truth*, billed as a sequel to *Pragmatism*. James starts by seemingly endorsing a common-sense correspondence view (ibid., pp. 96–7):

> Truth, as any dictionary will tell you, is a property of certain of our ideas. It means their 'agreement,' as falsity means their disagreement, with 'reality.' Pragmatists and intellectualists both accept this definition as a matter of course.

But, James continues, they 'begin to quarrel' when 'the question is raised as to what may precisely be meant by the term "agreement," and what by the term "reality," when reality is taken as something for our ideas to agree with'. James rejects the idea that our ideas should 'copy' reality. Instead, 'agreement with reality' is spelled out in terms of 'the truth's cash-value in experiential terms'.

This is just pragmatism asking 'its usual question': 'Grant an idea or belief to be true' – 'what concrete difference will its being true make in anyone's actual life?' James concludes:

> *True ideas are those that we can assimilate, validate, corroborate and verify. False ideas are those that we cannot.* That is the practical difference it makes to us to have true ideas; that, therefore, is the meaning of truth, for it is all that truth is known-as. (ibid., p. 97; emphasis original)

James makes various attempts to express what is involved in the pragmatic validation of ideas, and there are three important points. First, truth is a matter of connection between experiences. James stresses that the apparent sense of 'agreement' exhibited in a true idea is to be found in productive connections between different experiences that it enables us to form: *'ideas (which themselves are but parts of our experience) become true just in so far as they help us to get into satisfactory relation with other parts of our experience*, to summarize them and get about among them by conceptual short-cuts' (ibid., p. 34; emphasis original). James says that the main function of true thoughts is their 'go-between function' (ibid., p. 37).[26] Second, a truth needs to incorporate previously known truths: 'our theory must mediate between all previous truths and certain new experiences' (ibid., p. 104). And 'a new opinion counts as "true" just in proportion as it gratifies the individual's desire to assimilate the novel in his experience to his beliefs in stock. It must both lean on old truth and grasp new fact' (ibid., p. 36). Third, truth has a 'leading' or 'guiding' function:

> To 'agree' in the widest sense with a reality, *can only mean to be guided either straight up to it or into its surroundings, or to be put into such working touch with it as to handle either it or something connected with it better than if we disagreed.* Better either intellectually or practically! (ibid., p. 102; emphasis original)

The sense of being 'guided to' a reality is obscure. Being put into 'working touch with it' is clearer, as James explains further:

> Any idea that helps us to *deal*, whether practically or intellectually, with either the reality or its belongings, that doesn't entangle our progress in frustrations, that *fits*, in fact, and adapts our life to the reality's whole setting, will agree sufficiently to meet the requirement. It will hold true of that reality. (Ibid.; emphases original)

---

[26] Recognizing this point, we can also make better sense of some mysterious statement such as this: 'Truths emerge from facts; but they dip forward into facts again and add to them; which facts again create or reveal new truth (the word is indifferent) and so on indefinitely' (James [1907] 1975, p. 108).

Pragmatist Truth as Truth-by-Operational-Coherence

Truth-by-operational-coherence is something achievable and verifiable in practice to various degrees, and its pursuit is clearly useful. So it is certainly a notion of truth fit for pragmatist philosophy. And I believe that reinterpretation in terms of my ideas can give added clarity and plausibility to the classic pragmatist views, especially to James's view. Let me note some instructive points of comparison and contrast.

James's expressions can be blurry, sometimes verging on the poetic. This is the case, too, when he tries to explain what the practical bearings of truth are. I developed the notion of operational coherence precisely for the purpose of making James's kind of intuition more precise and systematic. James holds that pragmatism 'converts the absolutely empty notion of a static relation of "correspondence" ... between our minds and reality, into that of a rich and active commerce (that anyone may follow in detail and understand) between particular thoughts of ours, and the great universe of other experiences in which they play their parts and have their uses' (ibid., p. 39). James's idea of 'expediency' embodies the coercion of reality on thought, as explained in Section 1.6, and operational coherence does the same job. Both notions should be taken in a strong empiricist vein. (I think Dewey's notion of 'warranted assertibility' can also plausibly be incorporated into my notion of truth-by-operational-coherence.)

Recognizing truth as a quality rather than a black-and-white binary judgement, as I have explained in Section 4.4, can help shut down the ridicule that critics have heaped on the pragmatist attribution of truth to imperfect ideas. James's irritated response is understandable ('expedient in the long run, of course'), and so is the temptation felt by Peirce to reserve the designation of truth to the 'indefeasible'. But with my notion of truth we do not need to wait for the long run, and we can certainly refrain from Peirce's self-defeating move of appealing to the 'end of inquiry'. James easily recognized the qualitative and multi-dimensional nature of truth. Defining truth in terms of operational coherence also helps the pragmatist by taking truth one step away from direct verification in terms of success. This may be unsatisfying at first glance, but it does make it possible not to grant truth to accidental successes and failures due to variations of fringe circumstances.

I also think that an explicit adoption of pluralism, at both levels, can enrich and ease the pragmatist view on truth. First, adopting pluralism concerning the concept of truth (Section 4.2) liberates the pragmatist theory of truth from having to account for all uses of 'truth' by means of one idea. Identifying pragmatist truth with truth-by-operational coherence does not mean giving the same treatment to other kinds of truth. I think James ([1907] 1975, p. 34) went too far when he stated that the pragmatist account of truth was about 'what truth everywhere signifies' (if that meant unsuspecting participation in truth monism). Secondly, I think that losing a monist presumption concerning truth-by-operational-coherence itself (Section 4.5) can help the pragmatist view reach its full potential. I find Christopher Hookway's interpretation of Peirce instructive. Hookway (2004, p. 129) argues that Peirce's early convergentist view of truth was formulated so as to prop up a monist view of mind-independent reality (defining reality as the object of a true proposition), and that this view of truth was no longer necessary when Peirce's notion of reality moved on to an idea of direct perception. Hookway (ibid., p. 130) goes on to give a broader reading of what Peirce meant by any investigator being 'fated' to arrive at the truth: 'If a proposition is true, then anyone who investigates some question to which that proposition provides the answer is fated to believe it.' Hookway notes that this is 'compatible with rejection of the absolute conception of reality, for it is compatible with the view that our different perspectives are reflected in the varying ranges of questions that we can understand or take seriously'. Whether or not Hookway's view is the best interpretation of Peirce, I think it is a very plausible view on truth in its own right, consonant with the general spirit of pragmatism and very compatible with the plurality of truth-by-operational-coherence.

Beyond Misunderstanding

Having gone through the above presentation and interpretation, I think we can fully appreciate James's complaint about how the pragmatist view of truth 'suffered a hailstorm of contempt and ridicule' (James [1907] 1975, p. 38), 'so ferociously attacked by rationalistic philosophers, and so abominably misunderstood' (ibid., p. 95). He issued this complaint already as part of his best-known exposition of pragmatism in 1907 in defence of Dewey and Schiller as well as himself, but his attempts at clarification apparently had no effect in preventing the same kind of misunderstanding from recurring.

One factor must be noted, and set aside. Much vitriol directed against James's view on truth was excited by his willingness to entertain the possibility that religious belief could be justified as a pragmatic truth, if it leads one to live a good life. Russell (1910, p. 141) made no secret of the fact that this was his main objection to the pragmatist view of truth: 'It is chiefly in regard to religion that the pragmatist use of "truth" seems to me misleading.' For some philosophers this must have been the ultimate sign that there was something rotten in the pragmatist notion of truth, because they *knew* that God did not exist.

But as James was keen to emphasize, atheist dogmatism is just as unwarranted as theist dogmatism. In terms of my own notion of truth-by-operational-coherence, I admit that many coherent activities are carried out by many people on the basis of their belief in God, so I would be happy to accept that 'God exists' is true, to that extent. But such qualified truth would probably be rejected by the theists themselves: no, if God exists, he does so absolutely. Then we are asking if every activity in life can be carried out coherently by relying of the existence of God, and I think that is not the case – certainly not if we are talking about a bearded white man who lives in a place up there called heaven and listens to everyone's prayers, and even if we are talking more sensibly about an ultimate being who is omnipotent and supremely benevolent yet somehow allows all sorts of evil to take place in human life. But if it were the case (and it is a big 'if') that some societies could work out a truly all-encompassing and coherent way of life based on a well-crafted notion of a deity, then who are we to presume that such a God *doesn't* exist? Why would belief in God in that case be any less credible than the modern atheists' widespread belief in a rather extreme sort of materialism?

John Capps (2019, sec. 5) gives a convenient list of standard objections that have been raised against pragmatist theories of truth. I will examine each of objection and point out how I think it is based on a misunderstanding. First, Capps points out: 'if the pragmatic theory of truth equates truth with utility, this definition is (obviously!) refuted by the existence of useful but false beliefs, on the one hand, and by the existence of true but useless beliefs on the other'. Paul Horwich (1998b, p. 9) also takes the pragmatist to offer a straightforward definition of truth in terms of utility: 'Here truth is *utility*; true assumptions are those that work best – those which provoke actions with desirable results.' And some have even taken utility as happiness or psychological satisfaction, which is evidently not the main thing that was meant by the pragmatists. On this point I think it is enough to hear the lament from James ([1907] 1975, p. 111): 'A favorite formula for describing Mr. Schiller's doctrines and mine is that we are persons who think that by saying whatever you find it pleasant to say and calling it truth you fulfil every pragmatistic [sic] requirement.' Even worse is the interpretation of utility as material benefit. I am perhaps not alone in having been confronted by otherwise serious and well-informed scholars denouncing pragmatism as an American obsession with money-making, by taking James's talk of 'cash-value' (and what 'pays') quite literally!

If the first objection chides pragmatism for giving a wrong definition of truth, the second one charges it with confusion, of mistaking epistemic criteria for truth as the definition of truth. Here is how Capps expresses the objection: 'utility, long-term durability, and assertibility (etc.) should be viewed not as definitions but rather as criteria of truth.' This objection arises from not understanding the advantages of the pragmatists' semantic proposal concerning the meaning of 'truth', which I have tried to explain in Sections 4.1 and 4.3. It was for a reason that James gave the title of *The Meaning of Truth* to his sequel to *Pragmatism*, with an emphasis on 'meaning'. Russell did understand the pragmatists' semantic move, but gave a distorted view of it (1910, p. 138; emphases original): 'The arguments of pragmatists are almost wholly directed to proving that utility is a *criterion*; that utility is the *meaning* of truth is then supposed to follow.'[27] In my reading, at least James tackled the meaning of truth directly, rather than confusing criteria and definitions.

Thirdly on Capps's list, it is alleged that 'assessing the usefulness (etc.) of a belief is no more clear-cut than assessing its truth', which means that pragmatism fails to turn truth into a concept any more accessible and useful than the correspondence notion. Russell says: 'it is so often harder to determine whether a belief is useful than whether it is true' (Russell 1910, p. 138; see also p. 135). This objection draws a false equivalence between the in-principle inaccessiblity of the correspondence between idea and world, and the detailed vagueness and multi-dimensionality of very tangible successes, which I discussed in Section 4.4. The messiness of determining what is useful is only a problem if one has a monist notion of usefulness.

And finally, there is what Capps calls 'the fundamental objection': 'pragmatic theories of truth are anti-realist and, as such, violate basic intuitions about the nature and meaning of truth'. It is difficult to know where to start in response to this. First of all, the objection is question-begging, as it presumes to know what the right intuitions about truth are. But there is also a misunderstanding at a deeper level, as it ignores the realist dimension of pragmatism, in terms of how pragmatism demands that our ideas answer to experience, and to realities in the sense I defined them in Chapter 3. But in order to explain that point fully, I need to give a full exposition of pragmatist realism, which will be done in Chapter 5.

---

[27] And Russell resorts to ridicule instead of analysis: 'According to the pragmatists, to say "it is true that other people exist" *means* "it is useful to believe that other people exist"' (Russell 1910, p. 136; emphasis original).

CHAPTER 5

Realism

5.1 Overview

Why Do I Claim to Be a Realist?

I promised to offer a 'realism for realistic people' in this book, and I am now ready to spell out what this involves, having re-examined the nature of knowledge, reality and truth. What is realism, and what do I mean by making it realistic? Let us start by taking a step back: why should we care at all about something that might usefully be called 'realism', concerning science and various other practices in life? It is because we want to respect facts, not fantasies. Because we want to be open to learning from experience, instead of hiding behind the comforts of fixed opinions. Because we appreciate that empirical knowledge can help us live better in the material world. And because we want to live with the optimism that we *can* learn some truths about realities, rather than resign ourselves to living without improving our current state of ignorance.

Many people who have seen previous versions of my thoughts expressed in this book have cast doubt on the wisdom of calling my position 'realism' at all. There are different reasons for this. Some say that my position does not *deserve* to be called 'realism', because what I am calling 'reality' and 'truth' are too bound up with our human activities and concepts to be objects of anything called realism. But I think that is an unnecessary worry. As I emphasized in Section 2.1, entities being mind-framed does not imply that they are mind-controlled; there is no reason to be afraid to call them realities. Even though real entities are concept-bound, they do not obey our wishes. We can design a concept as we wish, but whether our concept can facilitate coherent activities is a matter that is quite outside our control. So I think my position does retain something very important in what many people value in realism. And my position is also clearly different from positivist or instrumentalist anti-realism, since I do think that theoretical propositions can be truth-apt. It is also different from van Fraassen's constructive empiricism, since I think that science can often learn truths about unobservable entities.

Identifying my position as realism involves a reconceptualization of the relation between realism and empiricism. In the philosophy of science the two are often perceived as opposing doctrines, and I think that this is a significant mistake. Realism should not mean anything contradictory to learning from experience, which is the essence of empiricism. I suspect that the perceived opposition between realism and empiricism is a historical accident. It is a legacy of the philosophical landscape of the early twentieth century, in which logical positivism was the standard-bearer of empiricism. The positivists on the whole shunned realism (with Neurath as a notable exception), as they saw it as part of the metaphysical tradition that they were trying to purge from philosophy. But as I have tried to show, realism concerning science and other practices does not need to rest on metaphysical realism. And empiricism is very much in line with the spirit of being realistic: if we stick with learning from experience, then we will not be getting into overblown claims or unattainable dreams of knowledge.

Some friendlier critics have advised against calling my position 'realism' for a different reason: the term has already been spoiled, as it were, with common meanings attached to it that I should not want. Better to leave this word behind, they say, and focus on articulating my ideas under some other name. However, there are strong motivations for keeping the 'realism' label. The same goes for the terms 'reality' and 'truth', which I have also proposed reconceptualizing in a pragmatist manner rather than setting aside altogether. Such powerful words should not be left to people who will use them in unhelpful ways. To make an extreme parallel: if those in power commit atrocities in the name of 'law and order', it is not the right response for the rest of us to retreat and stop insisting on the rule of law and an orderly society. The right response is to reclaim the terms, and use them to designate actions and situations that deserve to be described by those honorific words. We say *this* is order, and *that* is not the rule of law. Thankfully the current situation in philosophy is not so politically and morally charged, yet the same methodological point applies here, too. It would be a grave mistake to say that we don't care about truth, knowledge, reality and realism, just because these terms have been misused by others. So a certain amount of wrangling about what the important terms should mean is necessary. Recall how Peirce achieved nothing by trying to distinguish himself from other pragmatists by calling his own view 'pragmaticism', a label that no one took up (and he did not even expect people to take up). Dewey's move to talk about 'warranted assertibility' instead of truth was similarly unwise.

Being Realistic about Realism

It is an important part of my pragmatist stance that I think philosophy should play a role in guiding action. So I suggest that realism should be reconceived as a doctrine that we can actually put into practice, as an outlook on life that can guide the development of empirical knowledge in all walks of life. Being a realist or not should make some difference to what people do. Here we should heed Arthur Fine's (1984, p. 97) observation that everyone lives and works on the basis of what he calls the 'natural ontological attitude (NOA)', and philosophical realism layered on top of that only amounts to adding a 'desk-thumping, foot-stamping shout of "really!"' about what is already accepted. This is why we should set aside metaphysical realism (see Section 2.2): not because metaphysics is entirely meaningless as the positivists maintained, but because it doesn't change anything in practice, while disputes about it take attention away from more important issues. To the extent that 'scientific realism' is based on metaphysical realism, it also runs the same risk of futility. As I will discuss further in Section 5.2, I follow in the footsteps of various philosophers who have attempted to articulate productive versions of realism in a pragmatist mode, including Torretti, Vihalemm, Pihlström and Kitcher.

Now, if realism is to provide a guide to action, then it should be *realistic*, in the normal everyday sense of that word: 'having or showing a sensible and practical idea of what can be achieved or expected' (Google Dictionary, first definition). This also expresses a down-to-earth sense of rationality: to pursue aims that we can have some hope of achieving, at least aims that we can meaningfully work towards. Realism should be about the kind of knowledge that we can actually try to attain. As discussed in Chapter 2, standard scientific realism concerns itself with a kind of truth that we cannot meaningfully pursue, since we do not even know how to tell whether we are getting closer to truth as conceived in the traditional form of the correspondence theory. In contrast, realism as I intend it is not the search for an ultimate kind of truth. Rather, it is the pursuit of operational kinds of truth – namely, truth-by-operational-coherence, and secondary truth based on that (as explained in Chapter 4). These are the kinds of truth we can attain in the here-and-now; we can work on improving the truths we have, and we can clearly tell when we are making progress in that task. The same spirit holds for my notion of reality explained in Chapter 3. The term 'reality' should be reserved for things that we can meaningfully interact with, not for some inaccessible realm of Being that we only entertain in our abstract thought. My concepts make truth about real entities a very realistic aim to achieve, dissolving a central difficulty concerning standard scientific realism.

Several philosophers of science have noted the irony that many who call themselves 'realists' (subscribing to standard scientific realism) are not very realistic in their realism. Either they are wildly optimistic about what science has achieved or can achieve, or they talk at an incredible level of abstraction without any sense of what happens in real scientific research. In the Introduction I have already mentioned Peter Kosso's call for rendering realism more realistic, but he is by no means alone.[1] The 2011 workshop on practical realism in Tartu was subtitled 'Towards a Realistic Account of Science' (Lõhkivi and Vihalemm 2012, p. 1). Kerry McKenzie (2018) advocates 'being realistic' in the realism debate, seeking a kind of realist attitude that we can apply to science here and now. Kitcher's (2012, ch. 3) 'real realism' also seeks to ground realist reasoning in the knowledge of actually accessible objects. Teller's (2001) argument against the 'perfect model model' of science is also very much in the spirit of philosophically engaging with scientific knowledge in a form that humans can realistically pursue and attain. Putnam concludes his 'Defense of Internal Realism' thus: 'It is my view that reviving and revitalizing the realistic spirit is the important task for a philosopher at this time' (Putnam 1990c, p. 42).[2] This realistic spirit has been traced back to Wittgenstein by Cora Diamond (1991) and to Ramsey by Cheryl Misak (2020, pp. xxvi, 387).

There are two important ways in which the realistic spirit manifests itself in the context of debates on realism. First, it is often expressed as a kind of 'internalist' stance, which amounts to a commitment to deal with knowledge in terms of what is accessible to the knower. I think internalism is productive, though I would want to avoid the exclusive focus on conscious elements of an individual mind that is usually found in internalism in epistemology and the philosophy of mind. The internalist point I want to stress is that realism should offer operational guides to actual practice, rather than putting up inaccessible 'external' criteria. In this context Putnam's 'internal realism' is a very important cognate position

[1] Uskali Mäki (2009) advocates 'realistic realism', but with a rather different intention, namely that of curbing an excessive demand for literal truth in models in economics.
[2] It is not entirely clear exactly what Putnam meant by this, but David Macarthur (2012) takes it as a matter of respecting common sense, which remained an important part of Putnam's philosophical outlook.

to mine, as I will discuss further in Section 5.3. There I will also relate Putnam's position to a long tradition of seeing knowledge as situated in the knower, reaching ultimately back to Kant and currently very much alive in perspectivism as advocated by Ronald Giere, Michela Massimi and others.

The other manifestation of the realistic spirit is progress in an *iterative* form. This is because any inquiry, to be realistic, needs to start from a situation that we inherit from our personal and collective pasts. And our starting point is hardly ever going to be anything like a firm foundation – recall Peirce and Dewey pointing out that inquiry begins with a state of *dis*order. But if we don't have a firm foundation, where can knowledge rest? (One answer, of course, is that knowledge does not *rest* – or certainly *inquiry* does not.) As we try to forge ahead in inquiry, all we can do is to accept *something* in our inheritance as given, and learn what we can learn on that basis. And if we are fortunate enough, we will be able to use what we learn in order to come back to our starting point and improve it. I have used the term **epistemic iteration** to designate such a process of 'getting on' in the absence of indubitable foundations. In epistemic iteration successive stages of knowledge are created, each building on the preceding one, in order to enhance the achievement of certain epistemic goals. In each step, the later stage is based on the earlier stage but cannot be deduced from it; later developments often correct and refine the presuppositions made in the earlier stage. Initially I crafted this notion in the context of trying to solve the puzzle of how measurement standards can be established in the absence of pre-existing standards against which they might be validated (Chang 2004, ch. 5, and also pp. 44–8). By now I am quite confident that epistemic iteration is a very general feature of empirical inquiry in many different types of situation (see Chang 2007, 2016a, 2017a). I will give a more in-depth consideration of the iterative nature of realistic scientific progress in Section 5.5.

Activist Realism

Realism as I take it is not a descriptive thesis, but a commitment to an ideal. In ordinary usage an 'ism' might be a political ideology (as in 'nationalism', 'conservatism' or 'environmentalism') or an artistic movement (as in 'cubism' or 'impressionism').[3] It ought to be similar in philosophy: for example, positivism as practised by the Vienna Circle was certainly a committed stance, as is reductionism in the hands of

---

[3] More broadly it can designate a discernible stance or tendency of any kind: 'nihilism', 'nudism', 'egoism', 'narcissism' or even 'rheumatism'.

practising scientists. The core of real*ism* as I see it is an *activist ideal of inquiry*: a commitment to seek more and better knowledge about realities, along with a commitment to improve our epistemic practices to that end. The overall objective of realism ought to be the improvement of knowledge in its extent and quality, not only propositional knowledge but all aspects of active knowledge as characterized in Chapter 1. The realism I defend involves not only a recognition of realities, but a strenuous commitment to promote inquiry and facilitate learning. Realism concerning science[4] should be about how we should pursue scientific knowledge.

In previous work (Chang 2012a, p. 205) I called this ideal of inquiry 'active (scientific) realism': science should strive to maximize our contact with realities and our learning about them. To avoid conflation with the terminology of 'active' knowledge adopted in this book, I now propose to rename my brand of realism slightly, as **activist realism**.[5] The 'activist' label accentuates a normative outlook, focused on an imperative of progress. In this regard, my view on realism has not changed much from what was expressed a decade ago (ibid., ch. 4, esp. pp. 215–24), though I hope it has become better founded now, with a more detailed articulation of key concepts.

Activist realism is a commitment to do whatever we can in order to extend and enhance our knowledge concerning realities, as much as possible in the context of other aims and values. It is not an attitude of sitting content in appreciation of the knowledge that we have already attained, or condemning people who do not accept and use our knowledge. Activist realism means not only accepting the verdict of empirical tests of hypotheses, but also devising more and better tests, and producing more hypotheses to be tested. Activist realism also dictates that we should ask entirely new questions, make new theories, and even create more real entities and learn about them.

Like all good ideals, the activist-realist ideal I am advocating here may acquire an obvious ('duh!') quality to it, once it is seriously entertained: who would not want more knowledge about more realities? But it is not a trivial assumption that knowledge is a good thing; I would certainly not want to argue that knowledge is an absolute good that trumps all others, which is why the statement of activist realism only says that we should

---

[4] 'Scientific realism' would have been a good phrase for designating my position, of course, except that the phrase was taken long ago by people who mean something quite different. See Section 2.4.
[5] It is possible to think about active knowledge (knowledge-as-ability) without much interest in increasing it, in which case one would be non-activist about active knowledge.

pursue knowledge as much as possible in the context of other aims and values. With that qualification, however, I do think that activist realism is an ideal that should have a very broad appeal. Even those who identify as anti-realists in relation to standard scientific realism should be able to join in with activist realism; there are hardly any anti-realists who would argue against learning what we *can* actually learn. Rather, their objection to standard scientific realism is that it attributes to science the kind of knowledge that we cannot reasonably claim to be able to attain. In Section 5.6 I will show how activist realism relates to various positions in the debate on scientific realism, and hopefully prompt a productive realignment of positions in the realism debate.

Still, one might doubt that the articulation of activist realism would make any difference to scientific practice – aren't scientists already doing their utmost to increase knowledge? Shouldn't we simply leave it up to the scientists, to do what they do so well? What is the role of philosophical thinking in relation to this realism concerning science? But scientists do not always behave in a way that is maximally conducive to learning. Other thoughtful people can suggest, respectfully yet critically, how science might be done even better, while acknowledging that it is already done very well indeed. If philosophers of science do not perform this function, who will? I find great inspiration from a neighbouring discipline: Paul Forman (1991, p. 86) urges historians of science to embrace 'the obligation to decide for ourselves what is the good of science, and by our historical research and writing to advance that good'. The same goes for the function of the philosophy of science even more urgently.

Scientists often refuse to accept empirical evidence, or they explain it away by means of ad hoc hypotheses, while shutting down questions and conceptual schemes that offer alternatives to the currently dominant one. Such practices are anti-realist in my sense, since they constitute attempts to avoid full inquisitive engagement with realities. Any number of examples of such anti-realist practices can be found from the history of science, ranging from the early modern suppression of Copernicanism to the initial hostility to the ideas of prions and epigenetic inheritance in the late twentieth century. Sometimes scientists display a lack of concern that science should deal in concepts that facilitate empirical learning. We have seen theories shift in the direction of speculativeness, with fewer opportunities for empirical tests. Often the turn towards untestable hypotheses is motivated by metaphysical desiderata, such as unity and beauty; a lasting tradition of such work runs through the history of physics, from the ancient theory of four elements through the eighteenth-century system of imponderable fluids to the superstring theories of today.[6]

Activist realism can make a difference in practice, as one should expect from a pragmatist doctrine. The sense of futility about the realism debate (as expressed by Fine, quoted above) is only apt in connection with metaphysical realism or standard scientific realism. In relation to those realist doctrines, it can indeed be the case that a clash between philosophical realism and anti-realism makes very little difference in practice. For example, in much of nineteenth-century organic chemistry those who professed their belief in the reality of atoms seem to have done basically the same kind of work as those who didn't. Add an appropriate suspension of belief or disbelief, and the practices come out looking much the same on both sides. (A shouting match between scientists about realism is no more productive than one between philosophers.)

One particular debate that would not be productive to engage in is the one between realism and constructivism. Activist realism recognizes concepts and theories as human constructions. And according to my view of reality (see Chapter 3), the referents of conceptual constructions that facilitate operationally coherent activities *are* real. This blending of realism and constructivism should not worry anyone. We may propose the initial construction of a concept freely, but whether coherent activities can be carried out on the basis of it is not at all up to our will. In a well-executed process of inquiry there will be an iterative process of concept-building that responds to empirical successes and failures, akin to the process of resistance-and-accommodation that Pickering (1995) identifies in scientific practice. We do construct realities, but that process of construction is not arbitrary. Seeing realism and constructivism thus going together helps us make sense of an apparently constructivist statement from the staunch realist Ludwig Boltzmann: 'According to my feeling, the task of theory lies in the construction of an image of the exterior world, which exists only in our mind and should serve to guide us in all our thoughts and all our experiments' (quoted in Nye 1972, p. 20).

Activist Realism and Scientific Progress

Activist realism is an inherently progressivist doctrine, as it is a commitment to attaining more and better knowledge. So, in order to make

---

[6] See Hossenfelder (2018) for a critique of recent theoretical physics in this regard, and Dawid (2013) for a nuanced defence of string theory.

a full articulation of activist realism, I must make a careful consideration of the idea of progress in science. Progressivism, in the basic sense of 'wanting to make things better', has been a powerful modern ideal ever since the Age of Enlightenment. The advancement of science has been a central part of the Enlightenment vision, and progress has been an unquestioned credo within science itself. Although the desirability of progress is nearly tautological (how can making something better not be a good thing?), we should recognize that there are also some worthy non-progressivist ideals, such as the conservative yearning for order and stability, or the acceptance and resignation counselled by Buddhism. Even when it comes to knowledge, one may feel that we already know enough to support a decent kind of life, and that learning more is only likely to cause trouble. Whether this may be the correct attitude would need to be judged on a case-by-case basis. But on the whole I don't think that humanity is at a point at which further knowledge should be generally refused. Even though there are specific fronts on which a headlong dash to learn more is unwise and needs to be regulated through a careful consideration of likely consequences, such local restraint is compatible with an overall imperative of progress.

Progress was a central issue in the philosophy of science in the 1960s and the 1970s, but it has taken a back seat nowadays. Perhaps it is considered an outmoded modernist ideal by some. But scientific progress has been abandoned as a topic of discussion even by many who value it greatly, perhaps due to the lack of *philosophical progress* in giving a good characterization of scientific progress. But a lack of clear progress is not a good reason to abandon a philosophical question, and the neglect of progress in the philosophy of science is a grave mistake. It is difficult to deny that science does continue to progress, and we need to understand how it manages to do so, and in what sense exactly. It is ironic that many of the sophisticated thinkers who shun the talk of scientific progress nowadays are at the same time fervent and outspoken advocates of social and political progress. Although there is no logical contradiction in holding both of these attitudes, there is certainly discomfort. If the idea of 'progress' is so lacking in cogency as to be not worth considering even in the realm of science, how do we presume to have such a secure notion of it in the realm of ethics or politics? Moreover, science does and must continue to play a crucial role in social progress, both intellectually and materially. I endorse wholeheartedly Kitcher's sentiment, inherited from Dewey: progress has been a neglected concept lately, and we need to 'rehabilitate' this 'endangered concept', and restore its importance to inquiry.[7]

The considerations I have made in earlier chapters about knowledge, truth and inquiry strongly suggest that we should break away from the view of scientific progress that is found at the foundation of standard scientific realism, which sees progress as an approach to the Truth. Kitcher (forthcoming, ch. 2) presents 'pragmatic progress', which works as progress from a problematic situation in Peirce's and Dewey's spirit of inquiry, as a direct contrast to teleological progress. Activist realism adds another layer of motivation to move away from the teleological view of progress, and instead recommends a view of progress as an *abundant* form of development. The abundant shape of progress follows from the nature of inquiry as explained in Section 1.5. Inquiry is ultimately unrestricted, in the sense that any part of the problematic situation may be modified to resolve it. Different types of resolution point to different ways in which knowledge can be improved and science can progress.

Here I will only present a very brief (and not exhaustive) taxonomy of different ways of scientific progress, to give a sense of the abundance. Inspired by Kitcher's (1993, ch. 4) analysis of the 'varieties of progress', I give a non-exhaustive list here. (1) Given an existing question, we can answer it, or improve the quality of an existing answer. Improvements may involve correcting known isolated errors, or finding better methods of observation, measurement or computation. This will be a routine occurrence in the fact-gathering type of inquiry. (2) Within the realm of fact-gathering inquiry, we can also make progress by asking and answering new questions (Shan 2019). Asking new questions without discarding old ones allows an accumulation of propositional knowledge. (3) Not only can we make progress by asking new questions, but we can also improve the questions that we have been asking. Better questions seek information that is more relevant and useful to the achievement of the broader aims that we have. (4) Active knowledge can be improved by increasing the coherence of our activities. It happens routinely that we learn to execute a task more effectively, as anyone who has learned any skill can testify.[8] (5) Propositional knowledge can also be improved through the increase of coherence. Starting with a proposition that we accept, we can enhance its

---

[7] Kitcher (forthcoming, ch. 2); he has kept the issue of progress alive at least since Kitcher (1993).

[8] The first sequencing of the human genome was a painstaking enterprise that took thirteen years on a budget of about 1 billion dollars; nowadays it is done in a day or two for a few thousand dollars. Similarly for the manufacturing of an atomic bomb.

truth-by-operational-coherence by improving the coherence of the activities relying on it, or by devising more activities relying on it. (6) Active knowledge can be increased if we can find new (and preferably better) methods for achieving the same aim; multiple methods for achieving the same end may be maintained, and they may enhance each other. (7) When we create new material objects and phenomena, we can of course gain new knowledge about them, as well as the knowledge of how to make them. (8) We can create new concepts, and devise epistemic activities involving them. If these activities turn out to be coherent, then we have new realities, about which we can ask further questions. (9) We can set ourselves new aims, which we then learn how to achieve by devising new activities. If we are successful, we will certainly have acquired new active knowledge, and most likely some new propositional knowledge as well. (10) We can modify the aims of existing activities, so that they are more compatible with the aims of other activities we are engaged in, and they serve their external functions better. Such modifications increase the operational coherence of the systems of practice in which the activities in question occur.

So, if we ask 'what is scientific progress?', there will be no single answer that is informative enough. There are many different senses of scientific progress, ranging from the narrow and well prescribed to the completely open-ended. These different types of progress will be difficult to aggregate into one measure, and that is just fine. What is important is that we recognize all significant types of progress, so that our general notion of progress does not become impoverished or imbalanced. All of the modes of progress identified above should be, and are, pursued in science. It would be foolish to restrict our effort to only some of them, not to mention just one. When we consider all these multifarious ways of progress, Peirce's injunction comes truly alive: 'Do not block the way of inquiry.'[9] Rejecting the traditional view of scientific progress towards perfect truth, it would be more sensible to adopt the 'progress from' perspective advocated by Kuhn instead, inspired by an evolutionary model of scientific development. But if we are to think on the evolutionary template, it would perhaps be better to go with Rouse's (2015) picture of niche-construction, which fully acknowledges the dynamic and constructive interaction between the organism and its environment.

---

[9] This theme is central to Amy McLaughlin's interpretation of Peirce, articulated in McLaughlin (2009) and McLaughlin (2011); likewise for Susan Haack (2014).

Activist realism dictates a pluralistic shape of progress on a larger scale, too, as I will discuss in more detail in Section 5.4. In trying to gain knowledge in any way we can, we will most likely end up with multiple systems of practice in a given field of study. Each system will yield valuable knowledge in a distinctive way. In fact, my activist perspective has strong affinity to the very old-fashioned *cumulative* view of scientific progress, according to which we keep producing and storing up more and more knowledge (within each system of practice, and by the accumulation of systems). And we may create more knowledge yet, if we can establish meaningful links between the truths and realities found in different systems of practice. I am not making a logical deduction from activist realism to pluralism, but a cheerful and exuberant acceptance of the actual shape of the history of science and other types of inquiry. I am also recognizing that there is no compelling reason to expect that the future of inquiry will be radically different in this regard. Realism is generally taken as a monist position both in metaphysics and in the philosophy of science, but in my view monism must be rejected if it stands in the way of progress. Various successes of science will easily result in the establishment of various real entities and reveal various truths about each.

Such abundance of truth and reality should not embarrass us, as I have already argued in Sections 3.4 and 4.5. True realism will pursue knowledge unafraid in the spirit of Feyerabend, freed up from the unnecessary constraints of monism. All theories and all systems of practice that facilitate successful activities provide ways of learning, and they should all be maintained and developed actively. The picture of inquiry that emerges is that of the humble cultivation of abundance. Activist realism is a commitment to keep learning about the ever-multiplying realities in life. In order to maximize our learning about realities, we should continue to invent new concepts, while keeping all the old ones that can support coherent activities in variously separate and overlapping domains. This is how we acquire more knowledge about more realities.

These considerations on the nature of scientific progress also give us a fresh perspective on the state of the debate concerning scientific realism. Realism should be taken as a business of ampliative inquiry, not ampliative inference. Activist realism is not ampliative in the sense of drawing conclusions that are stronger than what the available evidence warrants, but in the sense of encouraging the actual creation of more knowledge. In science, all modes of inquiry should be put to work in this relentless drive to increase and improve knowledge. Bridgman (1955, p. 535) put it bluntly: 'The scientific method, as far as it is a method, is nothing more than doing one's damnedest with one's mind, no holds barred.' And rather than engaging in defensive manoeuvres attempting to show that science can approach an unattainable type of truth in some way, realism in philosophy should focus on articulating all the productive ways in which knowledge can grow.

5.2   Pragmatism and Realism

As already indicated, I see pragmatism as the most suitable philosophical tradition in which to base my conception of realism. Various other thinkers with cognate views to mine have also framed their ideas explicitly in pragmatist terms. I have been inspired by their attempts to show that realism can be conceived as a pragmatist position, and in this section I want to pay homage to these predecessors and draw more attention to their work. Roberto Torretti has explicitly advocated a 'pragmatic realism',[10] and very clearly pointed out what I have called the fallacy of pre-figuration. Rein Vihalemm and Endla Lôhkivi's 'practical realism' is a practice-based approach that has its roots in Marxist philosophy. Sami Pihlström's 'pragmatic scientific realism' is a notable synthesis of Kantianism and pragmatism, with affinities to C. I. Lewis's work. Philip Kitcher presents a revival of classical pragmatism in a way that is compatible with correspondence realism, which I think makes perfect sense if correspondence is taken as a relation that obtains in actual practice.

Roberto Torretti's Pragmatic Realism

Among my predecessors who have advocated placing realism within the pragmatist tradition, I cannot think of anyone more evocative than Roberto Torretti, whom I have quoted a few times already. Torretti (2000, p. 114) advocates 'pragmatic realism', which I have already acknowledged (Chang 2016b) as a strong inspiration for my thinking. At a very basic level, he sees realism and pragmatism as motivated by the same drive: realists are concerned about having correct knowledge of things; so are pragmatists. This is similar to my perception of affinity between realism and empiricism, and between empiricism and pragmatism. Torretti (2000, pp. 114–15) states that science is 'the continuation of common sense by other means', reflecting the insistence

---

[10] The phrase 'pragmatic realism', or something very close to it, is used by many authors, including Pihlström and Kitcher, whose works are also discussed in this section. I should also note Pickering (1995, p. 183) and Timothy Lenoir (1992, p. 166).

by Dewey and other pragmatists that the process of inquiry is continuous from everyday life to the most esoteric branches of science. In all cases, knowledge is about getting on in the world, dealing with our situation in the most effective way possible.

Yet, what philosophers commonly mean by (scientific) realism diverges quite strongly from pragmatism. Torretti suspects, correctly I think, that the kind of 'realism' stemming from what I have called the fallacy of pre-figuration (see Section 2.1) is a hangover from a monotheistic perspective on knowledge and scholarship: 'The existence of a well-defined or . . . ready-made reality is no doubt implied by the standard monotheistic conception of God, but I have not the slightest ground for thinking that God's worldview can be articulated in human discourse. To entertain the notion that we could convey that view in words is a symptom of acute provincialism' – even though this provincialism is so often dressed up as universalism! This is something that a mature philosophy of science ought to be able to transcend: 'it is pragmatic realism, not the nostalgic kryptotheology of "scientific realism", that best expresses the real facts of human knowledge and the working scientist's understanding of reality' (ibid., p. 115). Torretti maintains that 'science as it is actually practiced' is not concerned with looking for an 'absolute structure' of reality (ibid., p. 117).

The Tartu School of Practical Realism

My discussion in Chapter 2 benefited greatly from Rein Vihalemm's character-ization of standard scientific realism. Now it is time to pay attention to his own positive view on realism. Vihalemm, Endla Lõhkivi and their colleagues in Tartu have developed a position that they call 'practical realism' (see Lõhkivi and Vihalemm 2012 for a recent statement of their position, and a collection of papers by some of their most prominent fellow-travellers). Vihalemm does not quite identify his 'Marxist practice-based realism' with pragmatism, but I think the affinity with pragmatism is quite clear.[11] The main tenets of practical realism as given by Lõhkivi and Vihalemm (2012, p. 3) can be summarized as follows:

(1) Science does not represent the world 'as it really is' from a god's-eye point of view. It is not an ideal of science to pursue a 'one-to-one representation of reality', for which we do not have criteria of judgement.

---

[11] See Vihalemm (2012, p. 20), for phrase quoted here, and p. 11 for the distancing from pragmatism.

(2)   The inaccessiblity of the world independently of scientific theories, paradigms or practices does not argue for internal realism or 'radical' social constructivism.

(3)   Scientific research is a 'practical activity', whose main form is 'the scientific experiment that takes place in the real world, being a purposeful and critical theory-guided ... material interference with nature'.

(4)   'Science as practice is also a social-historical activity, which means ... that scientific practice includes a normative aspect, too.'

(5)   This position qualifies as realism, 'as it claims that ... science as practice is a way in which we are engaged with the world'.

I must say that I agree entirely with all of these tenets, at least if we understand the terms used in them (such as 'the world') in my pragmatist sense. As I see it, Vihalemm's fundamental idea behind practical realism is that all knowledge, or even all discourse, is rooted in practice:

> To speak about the world outside practice means to speak about something indefinable or illusory. It is only through practice that the objective world can really exist for humans. Therefore, knowledge must be regarded as the process of understanding how the world becomes defined in practice. (Vihalemm 2012, p. 10)

And what exactly is practice? It is 'human activity as a social-historical, critically purposeful-normative, constructive, material interference with nature and society producing and reproducing the human world – culture – in nature' (ibid.). Not only discourse, but objects themselves are grounded in practice. Vihalemm (ibid., p. 14) quotes Rouse (1987, p. 163) approvingly in this connection: 'Belonging to the realm of possible determinations open within our practices is constitutive of a thing's being a thing at all.' Vihalemm's metaphysics is a radical one:

> The practice-based approach implies that practical activity has a more fundamental status than the status of individual objects-things. Concrete determination of the existence of individual objects in this case is determined by specifically defined activities in the context of which these objects-things appear as specific invariants. (Vihalemm 2012, p. 13)

As his foil, Vihalemm identifies 'standard scientific realism', as discussed in Chapter 2. As for anti-realism, he defines it as any position that rejects one or more tenets of standard realism. He says that practical realism is opposed to both standard realism and instrumentalist/constructivist anti-realism. The main thing he finds wrong with both sides of the debate is that each position is 'isolated from practice' and 'does not proceed from the practice of real science' (ibid., p. 10).

Tenet 2 of practical realism is notable in its explicit disavowal of Putnam's internal realism, which I will discuss further in Section 5.3. Vihalemm regards internal realism as not worthy of the title 'realism', which is somewhat puzzling since his own position seems to share a great deal with internal realism.[12] He adds that an 'essential difference between internal realism and practical realism' is that internal realism 'belongs to the tradition of Kantianism and cannot actually be qualified as realism at all' (Vihalemm 2012, p. 17). Why Vihalemm objected so much to Kantianism is not entirely clear to me. Perhaps an important clue is given in his comment on the notion of truth (ibid., p. 19): 'I cannot speak for pragmatists, but in practical realism, "truth" can be interpreted in a deflationary way and this interpretation is compatible with semantic realism.' And Vihalemm traces semantic realism back to Niiniluoto's view that 'truth is a semantical relation between language and reality. Its meaning is given by a modern (Tarskian) version of the correspondence theory, and its best indicator is given by systematic enquiry using the methods of science' (Niiniluoto 1999, p. 10, quoted in Vihalemm 2012, p. 18). Like Putnam in his later phase, Vihalemm is pulled into disquotation and correspondence. But how is that compatible with the practice-based view of everything, which seemed to me very much like internalism? Vihalemm (ibid., p. 19) quotes Niiniluoto (1999, p. 11) who, in contrast to standard scientific realists, says that 'THE WORLD contains unidentified objects which are identifiable, but not "self-identifying objects" in the bad metaphysical sense'. What Vihalemm and Niiniluoto mean by 'THE WORLD' may seem like the Kantian noumenal world, but Vihalemm says that 'for practical realists or [Marxist] materialists it [the world] is not ungraspable, but identifiable in its concrete forms of existence through practice'. As Vihalemm (2012, p. 20) notes with approval, Niiniluoto (p. 275) quotes Friedrich Engels from his 1886 text on Feuerbach in this connection: 'If we are able to prove the correctness of our conception of a natural process by making it ourselves . . . then there is an end to the Kantian ungraspable "thing-in-itself".' But I am not sure whether being graspable through practice is really different from only being cognizable as phenomena inextricably wrapped in the Jamesian 'trail of the human serpent'.

---

[12] Lõhkivi thinks that this might have been due to a current of externalism in Vihalemm's thought (conversation in Tartu, March 2017).

Sami Pihlström's 'Pragmatic Scientific Realism'

Much as I find practical realism congenial, I do not find Vihalemm's treatment of Kant convincing. In my view, the best way to build on Kantian insights is shown by Sami Pihlström's exciting blend of pragmatism and Kantianism (see Pihlström 2014 for the latest synthesis). He is an avowed pragmatist (unlike Putnam and Vihalemm) and he regards realism as a crucially important problem to be tackled by philosophy (Pihlström 2011, p. 121), so it will be very important for me to see how he positions himself. According to my understanding, Pihlström's main insight on the realism debate is twofold: we need Kant, and what we need is a pragmatist version of Kant. To be more precise: Pihlström (2011, p. 111) wants to hold on to the fundamental Kantian take on the realism debate, which is a combination of transcendental idealism (cognition requires a conceptual scheme, which is indeed mind-dependent) and empirical realism (the empirical properties of objects, cognized within the conceptual scheme, are not controlled by the cognizing subject).[13] This is exactly how I see the situation: reality is mind-framed but not mind-controlled; we cannot claim any absolute validity for our human framing of cognition, but *within* that framing the objects we deal with are real. (Pihlström also speaks of 'scheme-internal' realism, which seems to be a nod to Carnap, and in good accord with Putnam's internal realism.)

In making a pragmatist rendition of the fundamental Kantian insight, Pihlström recognizes conceptual schemes as rooted in practice. Pihlström (2011, p. 112; also 2012, p. 85) characterizes this move as a sort of 'pragmatic "naturalization" of Kantian transcendental idealism'. Conceptual schemes are not universal and static as Kant had assumed; rather, as epistemic subjects 'we are fully naturally situated within context-dependent and context-changing practices'. It is these practices that 'contain "relative *a priori*" conditions that structure our ways of experiencing reality'. Pihlström's stance on this is quite consonant with Friedman's neo-Kantianism, and I think Pihlström is also correct in reading Kuhn along these lines, and in tracing the idea of the 'relative *a priori*' to C. I. Lewis (Pihlström 2012, pp. 82–3). What is shared by all these quasi-Kantian positions is the recognition that cognition is only possible on the presumption of certain a priori principles. They also share a *denial* of the Kantian insistence that the a priori principles to be presupposed are universal, fixed and inevitable. To put the point paradoxically: the a priori is necessary, but only contingently so (see Section 3.2).

---

[13] For non-experts, the exposition by Nicholas Stang (2018) is quite helpful on these basic Kantian doctrines involved here.

For Pihlström, ontology is rooted in practice: 'It is in our goal-directed activities and practices themselves that our ontological ways of taking the world to be in some particular manner are to be located.' This he identifies as consonant with Dewey's view that 'scientific objects are not "ready made" prior to inquiry but rather arise out of, or are constructed and/or identified in the course of, inquiry' (ibid., p. 89, paraphrasing Dewey). Furthermore, 'there are, and can be, no beliefs at all apart from such activities and practices' (ibid., p. 84). As Pihlström himself recognizes, these thoughts are strongly resonant with Vihalemm's view that truth and reality are only meaningful within practice, and similarly with Putnam's internal realist view. Here is one cogent summary of Pihlström's position (2012, p. 80): 'the Kantian or quasi-Kantian "transcendental" element – whatever it is that must be presupposed for inquiry, representation, or cognition to be possible – may lie in the local practices themselves'. And he points out that 'a very basic transcendental issue concerning the practice-laden representability and experienceability of reality must be taken up from the perspective of [Vihalemm's] practical realism, too' (ibid., p. 88).

I am mostly in enthusiastic agreement with Pihlström, and it seems to me that there is an important further step to take, after accepting his basic outlook: we must go into the details of specific practices and demonstrate, by some sort of transcendental argument, which specific assumptions are needed for which specific practices. We should also think again about the nature of transcendental arguments (see Chang 2008): what exactly is the method of sussing out the necessary enabling conditions of something? It is usually seen as a deduction, but it is not a straightforward kind of deduction – if it were, it wouldn't be such hard work following Kant's arguments! These are the lines of work that I have tried to carry out in my discussion of 'mind-framing' in Chapter 3.

Philip Kitcher's 'Hybrid Pragmatism' and 'Real Realism'

I cannot possibly close this discussion of philosophers of science currently pursuing the integration of pragmatism and realism without attention to the recent and ongoing work of Philip Kitcher. I have alluded to various aspects of his thought in the course of the earlier chapters, but this is an appropriate place to present his work in a more systematic way. Kitcher's brand of pragmatism has been laid out in various articles collected in the 2012 volume *Preludes to Pragmatism*, and in more recent works including the 2020 Descartes Lectures at Tilburg University, which will be synthesized into a forthcoming book entitled *Homo Quaerens: Progress, Truth, and Values*.

Building on the ideas advanced in the paper on 'Pragmatism and Realism: A Modest Proposal' (Kitcher 2012, ch. 5), his recent work explicitly advocates a 'hybrid pragmatism', at the core of which is a synthesis of the pragmatist and the correspondence theories of truth. Kitcher makes a new Peircian definition of truth: 'A sentence S of language L is true just in case, as used in L, that sentence would be stably retained as inquiry progresses indefinitely.' But he adds: 'Truth for descriptive sentences may be an amalgam, to which both the idea of correspondence and the Peircean exposition make necessary contributions' (Kitcher forthcoming, ch. 3). It is important for Kitcher to hold on to a notion of truth as correspondence, most of all because it serves the function of explaining pragmatic success (Kitcher 2012, ch. 4). Preserving the explanation of success by reference to truth is central to the version of scientific realism that Kitcher wants to preserve and defend. But 'real realism', as he puts it, is a down-to-earth doctrine, according to which the 'truth causes success' thesis is an empirical hypothesis, directly verifiable in some cases and to be trusted in other situations. What he calls the 'Galilean strategy' of real realism (realism in practice) is a piecemeal extension of that successful-because-true hypothesis from a well-confirmed domain to a new uncharted domain. Galileo's work with the telescope provides a persuasive illustration: the successful use of the telescope in terrestrial situations can be explained by the verifiable correspondence between how things look close-up to the naked eye, and how they look through a telescope from afar. In using the telescope to learn about astronomical objects, Galileo was conjecturing that its pragmatic success was also due to the correspondence of the telescopic images and the real shape of the objects, in this case not accessible to direct observation (Kitcher 2012, ch. 3; Kitcher forthcoming, ch. 4).

What I find most appealing in Kitcher's synthesis is the fact that he treats correspondence as something that takes place within the 'world of experience'; in that regard Kitcher's pragmatist realism is quite consonant with internal realism and perspectival realism (see Section 5.3). While he is with many metaphysical realists and standard scientific realists in invoking Tarski for the understanding of truth, he is also clear about a Tarskian point not usually emphasized by them: 'If (as I prefer) we want to apply truth to sentences, we have to make, or presuppose, reference to a language' (Kitcher forthcoming, ch. 3). And being language-bound means already having the affordances of thinking in terms of the entities that exist in the 'world of experience' of the speakers of a language, which are realities bound up with various primary truths expressed in that language. This is quite consonant with the view of mind-framed reality that I have proposed in Chapter 3.

Yet my view differs from Kitcher's in a few important respects. While he recognizes that correspondence is not correspondence to inaccessible noumenal reality, he does retain the traditional thought that empirical truth is always a matter of correspondence. Therefore, truth concerning objects that are not directly accessible to us can only be *modelled* on the notion of correspondence, and has to be treated as a hypothesis. In my view, it is more productive to allow truth-by-operational-coherence to play a role, for two main reasons. Correspondence truth is secondary truth, which only makes sense if there are previously established facts to which the propositions under consideration can correspond, and ultimately in the chain of correspondence there must be some primary truths. For the pragmatist the only primary truth one can rely on in empirical domains is truth-by-operational-coherence. Also, we should allow the possibility that a statement that is in principle able to attain correspondence truth may be functioning as truth-by-operational-coherence.

5.3 Internal and Perspectival Realism

In this section I examine two further positions on realism that are cognate to my own, especially in their embodiment of the realistic spirit. One is Putnam's internal realism, based on the insight that we need to make sense of ontology, truth and correspondence within given conceptual frameworks. I will also consider why Putnam renounced his internal realism, and suggest that a more strongly pragmatist interpretation of internal realism with the help of my notion of truth-as-operational-coherence would have made it more defensible. Closely related to internal realism is perspectivism (or, perspectival realism), which is also a species of internalism guided by the realistic spirit. I believe that the lasting lesson of perspectivism is the recognition that perspectival truth is something we can have and value, in line with truth-as-operational-coherence as I conceive it.

What Is Internal Realism?

Many philosophers over the ages have advanced positions that embody the realistic spirit. One notable tradition may be identified as epistemological internalism: the view that knowledge only exists as situated in actual knowers, recognizing the crucial importance of what I call the mind-framing of realities. Under various notions such as conceptual frameworks, paradigms and world-versions, internalists have recognized that knowledge can only exist within a specific system of practice. In the late twentieth century the most prominent expression of this view was Putnam's internal realism.

What exactly did Putnam mean by 'internal realism'? This in itself is a contentious issue, on which there is considerable commentary.[14] One of the reasons why it is difficult to give a straightforward characterization of Putnam's internal realism is that he defined it primarily in a negative way, as the opposite of 'external' or 'metaphysical' realism. In *Reason, Truth and History*,[15] he presents 'the perspective of metaphysical realism' as follows:

> On this perspective, the world consists of some fixed totality of mind-independent objects. There is exactly one true and complete description of 'the way the world is'. Truth involves some sort of correspondence relation between words or thought-signs and external things and sets of things. I shall call this perspective the *externalist* perspective, because its favorite point of view is a God's Eye point of view. (Putnam 1981, p. 49; emphasis original)[16]

In contrast, Putnam's own perspective is that both ontology and truth are 'internal' matters, only meaningful within a given conceptual framework. So, '*what objects does the world consist of?* is a question that it only makes sense to ask *within* a theory or description' (ibid.; emphases original). He adds: '"Objects" do not exist independently of conceptual schemes' (ibid., p. 52). And if ontology is internal, then it is inevitable that truth will be, too:

> If objects are . . . theory-dependent, then the whole idea of truth's being defined or explained in terms of a 'correspondence' between items in a language and items in a fixed theory-independent reality has to be given up. (Putnam 1990c, p. 41)

So what is truth for an internalist?

> 'Truth', in an internalist view, is some sort of (idealized) rational acceptability – some sort of ideal coherence of our beliefs with each other and with our

---

[14] See especially Hacking 1983; Steinhoff 1986; Sosa 1993; Clark and Hale 1994; Niiniluoto 1999; Baghramian 2012; and Button 2013. Putnam himself (2015a, p. 82) locates the first statement of 'internal realism' in his APA Eastern Division Presidential Address given in December 1976, published in *Meaning and the Moral Sciences* (1978), and says that the position was expounded further in *Reason, Truth and History* (1981). Putnam (1987) and Putnam (1990a) are also key texts.

[15] The exposition in this text is as clear and explicit a statement of internal realism as one can find in Putnam's own words, and it is taken as definitive by Niiniluoto (1999, p. 211) and Hacking (1983, pp. 92–3).

[16] Putnam (1990c, pp. 30–1) later notes that Field has distinguished the three theses as separate versions of metaphysical realism, and notes that 'the natural way of understanding' the second of these theses involves accepting the first one.

experiences *as those experiences are themselves represented in our belief system* – and not correspondence with mind-independent or discourse-independent 'states of affairs'. (Putnam 1981, pp. 49–50; emphasis original)

A fundamental affinity I find with internalism is its focus on what is accessible to us. If internalism is reconceived as a commitment to understanding truth and reality in terms of what we can *experience*, then it becomes seamlessly connected with empiricism and pragmatism. Niiniluoto (1999, p. 205) explains that Putnam's internal realism 'belongs to the tradition of Kantianism in its denial that the world has a "ready-made" structure, and to pragmatism in its linkage between truth and the epistemic concepts of verification and acceptance'. This blend of Kantianism and pragmatism, fully and explicitly developed by Pihlström (see Section 5.2), is also at the heart of my own take on realism. Although Putnam did not clearly identify himself as a pragmatist and called pragmatism an 'open question' (Putnam 1995, subtitle), I believe it is productive to make a pragmatist reading of Putnam. Interestingly, Torretti (2000, p. 114) cites Putnam's internal realism as a direct inspiration for his own 'pragmatic realism' (see Section 5.2), and notes that Putnam (1987, p. 17) stated later that he should have called his position 'pragmatic realism' instead of 'internal realism'.

In motivating and appreciating a pragmatist interpretation of Putnam, there are two key items to consider further: truth and correspondence. A pragmatist reading of Putnam's internal realism rests on taking 'experience' in a broad sense linked to actions and practices (see Section 1.6), not in the sense of information-input through perception. Read Putnam again with that in mind, when he says that truth consists in the 'coherence of our beliefs ... with our experiences'. With a fuller notion of experience, Putnam's notion of truth becomes quite close to my notion of truth-by-operational-coherence (Chapter 4). I believe that this reading is faithful to Putnam's own spirit, in wanting an operational notion of truth, rendering it an internal notion: 'All I ask is that what is supposed to be "true" be *warrantable* on the basis of experience and intelligence for creatures with a "rational and sensible nature"' (Putnam 1990c, p. 41). This impulse squares well with that of the pragmatists: 'What I believe is that there is *a* notion of truth, or, more humbly, of being "right," which we use constantly and which is not at all the metaphysical realist's notion of a description which "corresponds" to the noumenal facts' (ibid., p. 40). Even after he moved away from internal realism, Putnam maintained this orientation towards the concept of truth. Here is a pithy statement from his commentary on James: 'truth, James believes, must be such that we can say how it is possible for us to grasp what it is' (Putnam 1995, p. 10).

Putnam's internal-realist view on correspondence is also strikingly pragmatist, and very much like my own (see Section 2.5), so much so that I wonder if I had long ago absorbed it from him without realizing its importance. The main point is that correspondence is a perfectly operable notion within a system:

> [A] sign that is actually employed in a particular way by a particular community of users can correspond to particular objects *within the conceptual scheme of those users* ... *We* cut up the world into objects when we introduce one or another scheme of description. Since the objects *and* the signs are alike *internal* to the scheme of description, it is possible to say what matches what. (Putnam 1981, p. 52; emphases original)

Such passages clearly indicate Putnam's view that reference is an internal matter. And from this it is only a short step to recognize that there is a perfectly sensible internalist correspondence notion of truth, which I have called truth-by-comparison in Section 4.2. I prefer to distinguish truth-by-comparison, which is a matter of secondary truth, from truth-by-operational-coherence, which is primary. Putnam does not make that sort of distinction, but both of these types of truth are internalist in Putnam's sense.

Putnam's Arguments for Internal Realism

So much for what internal realism is and why I find it congenial. How did Putnam himself actually argue for this position? One line of argument begins with a commitment to common sense:

> If there is any appeal of Realism which is wholly legitimate it is the appeal to the commonsense feeling that *of course* there are tables and chairs, and any philosophy that tells us that there really aren't – that there are really only sense data, or only 'texts', or whatever, is more than slightly crazy. (Putnam 1987, pp. 3–4; emphasis original)

Putnam says that what is standardly called Realism is a betrayal of this insight: this Realism defeats Anti-realism by appealing to common sense, and then turns itself into 'Scientific Realism' that says everyday objects actually aren't real and only god-knows-what sanctioned in 'finished science' are real (ibid., p. 4). Rejecting that betrayal, Putnam opts for 'realism (with a small "r") ... that takes our familiar commonsense scheme, as well as our scientific and artistic and other schemes, at face value, without helping itself to the notion of the thing "in itself".' He declares: 'Realism with a capital "R" is, sad to say, the foe, not the defender, of realism with a small "r".' So we arrive at another brief statement on internal realism: it is 'the key to working out the programme of preserving commonsense realism while avoiding the absurdities and antinomies of metaphysical realism' (ibid., p. 17). These intuitions were in fact preserved even as Putnam renounced internal realism, as we shall see below.

But the argument for internal realism that got most attention in the literature is the 'model-theoretic argument', which is not so much an argument for internal realism but an argument against external/metaphysical realism. I will not try to retrace the technical intricacies of Putnam's argument(s), on which I would defer most of all to Tim Button's (2013) excellent and refined critical exposition. Here I only wish to offer an intuitive perspective on what useful lessons we can take from Putnam. Of the two classes of Putnam's model-theoretic argument that Button distinguishes, the more intuitively forceful is the class of 'indeterminacy arguments', which seek to demonstrate: '*If there is any way to make a theory true, then there are many ways to do so*' (Button 2013, p. 14; emphasis original). Putnam (1981, pp. 32–3) wants to demonstrate that the 'received view of interpretation', according to which the extensions and intensions of terms are fixed by fixing the truth-conditions of whole sentences, does not work. Extending Quine's insights, Putnam argues that 'it is possible to interpret the entire language in violently different ways, each of them compatible with the requirement that the truth-value of each sentence in each possible world would be the one specified'. In Hacking's homespun rendition, the thought goes like this: 'Every time you speak of cherries, you could be referring to what I call cats, and vice versa. Were I seriously to say that a cat is on a mat, you could assent, because you took me to be saying that a cherry is on a tree. We can reach total agreement on the facts of the world' (Hacking 1983, p. 102). Somewhat more abstractly, Button (2013, pp. 14–15) explains the 'permutation argument' (as the easiest of the model-theoretic arguments): 'Imagine that we were to lay out all the objects in the world, together with various labels (names) for them . . . Suppose we now shuffle the objects around. So long as we do not disturb the labels, exactly the same sentences will come out as true after the shuffling as were true before the shuffling.' This destroys the definite correspondence between words and objects that metaphysical realists presume as a fundamental tenet of their view.[17]

---

[17] Putnam's critique of metaphysical realism can appear self-defeating, because he sets up the model-theoretic arguments by first adopting the basic terms of metaphysical realism. Inherent in the very set-up of model theory, we have language L, domain X of individuals, and interpretation-function I from L to X (see Niiniluoto 1999, pp. 52ff.). This presupposes at the outset a world composed of well-defined individuals (without mind-framing). Putnam's model-theoretic arguments only make sense as a *reductio ad absurdum* of metaphysical realism, as Button suggests: 'Putnam does not

Why Did Putnam Move on from Internal Realism?

As I feel such a strong affinity to Putnam's internal realism, I must deal with one familiar issue concerning Putnam's philosophy: he often changed his mind. His view on realism was no exception, and from the early 1990s he explicitly disavowed internal realism. There is something uncomfortable, of course, about building on a philosophy that its author himself later renounced. Hear Putnam's own retrospective accounts of how he gave up internal realism (or, 'antirealism'!):

> I publicly renounced the thesis that true statements are those that we would accept were conditions to become sufficiently 'ideal,' which was the form of antirealism I defended in the late 1970s and '80s, as a mistaken 'concession to verificationism' ... (Putnam 2015b, p. 508, note 6)[18]

Here Putnam points to the pragmatist-leaning account of truth as the main thing that was wrong with his internal realism.

Interestingly, Putnam actually retained some crucial aspects of his internal realism after his explicit rejection of it. Even after renouncing internal realism, Putnam said that he remained committed to 'conceptual relativity', and I think he was right in maintaining that commitment. Putnam did not quite renounce pragmatism in general, either. On the contrary, he followed his renunciation of internal realism with a very sympathetic study of pragmatism (Putnam 1995). His 2015 retrospective says that what he did in 1990 was to return to what James had called 'natural realism', or 'direct realism' (Putnam 2015a, pp. 95–7), which was indeed how he framed his new attitude in *The Threefold Cord*, published in 1999. I think he also retained the realistic spirit (as discussed in Section 5.1), and this was part of his humanism: 'Our ideas of interpretation, explanation, and the rest flow as much from deep and complex human needs as our ethical values do' (Putnam 1990c, p. 37). All in all, I do not find that Putnam moved on to another convincing and coherent view after renouncing internal realism. Rather, I wish he would have retained internal realism and developed it further, and I think having a defensible pragmatist theory of truth would have helped him not lose his nerve about internal realism. I like to imagine that Putnam would have approved of my work. I believe that the notion of truth-by-operational-coherence allows a

---

embrace meaning scepticism; instead, he uses it as a *reductio* of opposing positions, such as external realism' (Button 2013, p. 3; emphasis original). Putnam (1977, p. 489) does say that he is showing how metaphysical realism 'collapses into incoherence'.

[18] See also Putnam (2015a, pp. 83–4, 91; 2015b, p. 502).

productive reinterpretation of both internal realism and pragmatism, and a synthesis of the two.

Perspectival Realism

What I see as the core of internal realism was present long before Putnam. As mentioned in Section 3.4, it was very much the spirit of Carnap's late work (see Carnap 1963, p. 871). It was strong in the works of pragmatist-leaning neo-Kantians before Putnam, including C. I. Lewis, Goodman and Kuhn. In current philosophy of science, this tradition continues under the name of perspectivism. Ronald Giere's conception of perspectivism (2006, pp. 13–14) starts metaphorically with 'the idea of viewing objects or scenes from different places'. But what he calls his 'prototype for a scientific perspectivism' is colour vision: 'colors are real enough, but … their reality is perspectival'. He proposes to understand perception in general this way, and also instrument-aided observation (ibid., ch. 3). And then comes the most 'controversial' and interesting extension of this idea, to scientific theorizing (ibid., ch. 4): 'the grand principles objectivists cite as universal laws of nature are better understood as defining highly generalized models that characterize a theoretical perspective' (ibid., p. 14). Building on Giere's work, Michela Massimi understands perspectivism as 'a family of positions that in different ways place emphasis on our *scientific knowledge being situated*' historically and culturally (Massimi 2018a, p. 164; emphasis original). For Teller (2018, p. 162) the key source of perspectivality is 'the different, even incompatible, modelling idealizations needed in practice for treatment of different aspects of a subject matter'.

Having heard that much, one might fairly ask a series of questions. What exactly is a perspective? What does situatedness consist in? And what exactly is it that gets situated? According to Massimi (2018b, p. 343, footnote 2), a perspective is 'the actual – historically and intellectually situated – scientific practice of any real scientific community at any given historical time'. What she means by 'practice' includes knowledge claims, methods, and norms of justification. In another place Massimi (2018a, p. 164) lists 'scientific representations, modelling practices, data gathering, and scientific theories' as the elements of knowledge that are situated. Put that way, a perspective is not so different from a Kuhnian paradigm in the sense of 'disciplinary matrix'; Giere (2006, p. 82) states that perspective is a narrower notion than paradigm, but does not elaborate on that point.

I think it is useful to distinguish three separate layers of perspectivality, which I laid out in a recent work comparing perspectivism and pragmatism (Chang 2020a; cf. Chakravartty 2010). (1) The same content can be expressed in different ways, in different languages, or using different expressions, that are *not* incommensurable with each other. The different expressions will typically have different connotations embodying divergent expectations and prompting divergent courses of action. For example, the Newtonian, Lagrangian and Hamiltonian formulations of classical mechanics are equivalent to each other in content, but with significantly differing affordances in problem-solving and further theorizing. (2) Different perspectives can highlight different aspects of a given object, and also conceal other aspects. This sense of perspectivism is consonant with quite a literal reading of 'perspective'. If we look at a three-dimensional object in the normal way, we will only see a two-dimensional picture whose content depends on the direction of gaze. A cylinder may look like a circle or a square, depending on the perspective one takes on it; the view of the circle conceals the square-ness of the other view,' and vice versa. Following Giere, we can generalize and extend this thought to both observational and theoretical perspectives. (3) Going deeper, one can argue that the relation between our knowledge and the world cannot be spelled out in an objectivist way. Any phenomena that we can discuss are already expressed in terms of concepts (mind-framed), and we can only choose from different conceptual frameworks that are liable to be incommensurable with each other. Even to say that two representations are different perspectives *on the same object* is to take too much for granted. Each perspective offers knowledge about realities, but not the same realities.

Both Giere and Massimi present their perspectivism as a realist position, a happy medium that recognizes both the situatedness of knowledge and the expression of mind-uncontrolled truth in situated knowledge. Perspectival realism allows us to transcend the opposition between 'objectivism (or objectivist realism)' and 'constructivism' (Giere 2006, p. 88). Massimi sees perspectival realism as 'the latest attempt at bypassing dichotomous divisions' in the realism debate: it is possible our 'scientific knowledge claims' are '*perspectival*, while also being claims about the world as *it is*' (Massimi 2018b, p. 342, emphases original). But how is that possible? Giere (2006, p. 81) stresses that perspectival facts are clearly facts; there is 'truth within a perspective' that is as robust as anything. To adapt one of his examples: looking north from Taiwan, Japan lies to the right of Korea; this is easily verifiable by experience, and quite indisputable, and entirely compatible with the fact that within the perspective from Kamchatka, Japan is to the left of Korea. And since any observing and theorizing must take place perspectivally, *all* truth claims are 'relative to a perspective'. But one can and should take a maximally realist attitude about such perspectival truth, in any reasonable sense of 'realist': in Giere's words again (2016, p. 138, emphasis original): 'claims made from within a perspective are nevertheless *intended* to be genuinely about the world, and thus "realistic," even though not fully precise or complete'. This notion of perspectival truth is very consonant with my notion of truth-by-operational-coherence, especially given Massimi's view that perspectives consist in sets of practices. And embracing the pragmatist notion of truth would push perspectivism into adopting the deepest sense of perspectivality (option 3 above).

But one may worry that we are now leaning too far in the direction of constructivism. Massimi pushes beyond Giere's argument, in a bid to show that perspectival truth can actually tell us something non-perspectival. Different perspectives may come to an agreement on specific points ('agreeing-whilst-perpsectivally-diagreeing') such as the charge of the electron as a minimal unit of electric charge (Massimi 2021). More generally, a productive interaction of different perspectives can point us to non-perspectival knowledge. But if the kind of scenario envisaged by Massimi worked out *too well*, it would provide fodder for the critique of perspectivism by Chakravartty, who argues that perspectivist arguments merely point to incomplete or idealized versions of fully objective and non-perspectival truth: 'one may speak the truth … without thereby speaking the whole truth … But this is not tantamount to perspectivism' (Chakravartty 2010, p. 407). Concerning the fact that scientists work with mutually inconsistent models of the same sets of phenomena or objects, Chakravartty contends (ibid., p. 406): 'even though there are thoroughly reasonable senses in which scientific models – and in particular, inconsistent models … are perspectival, this does not entail that we do not or cannot learn non-perspectival facts relating to the things these models model'. So if Massimi's triangulation from perspectival to non-perspectival knowledge is successful, that will only pave the way to Chakravartty's dismissal of perspectivism, which brings us right back to standard scientific realism.

5.4 Pluralism and Realism

The activist and realistic spirit of realism that I advocate recommends that we do whatever we plausibly can in order to enhance our knowledge of realities. Adopting realism as I take it enhances the arguments for pluralism advanced in Sections 3.4 and 4.5. Concerning pluralism more generally, I mostly stand by the view that I have expressed in previous publications, including the answers given there to worries about the plausibility and consequences of it. Here I will further develop some aspects of my thinking about pluralism that are enhanced by the articulation of realism made in this book.[19] First, the realistic spirit of internalism (see Section 5.3) naturally allows the plurality of conceptual schemes. Second, the realistic and the activist inclinations together provide strong support for what I have called 'conservationist pluralism', which takes care not to discard practices with good track records of success. Third, activist realism is best served by interactive pluralism, which seeks to reap benefits from productive interactions between different systems of practice, going beyond the non-interactive co-existence of different systems found in 'tolerant pluralism' or 'foliated pluralism'.

Internalism and Pluralism

Internal realism and other related positions discussed in Section 5.3 are strongly conducive to pluralism. The link between perspectivism and pluralism is obvious: a perspective on an object wouldn't be a *perspective* if it were impossible to take other perspectives on the same object. Similarly the internal–external distinction *almost* implies pluralism: if internally valid propositions are formulated within a conceptual scheme, then whole other sets of valid propositions will be formulated within other conceptual schemes. I say 'almost' because it is possible to insist, as Kant did, that there is only one fundamental conceptual scheme for cognition. But the whole histories of neo-Kantian philosophy, modern mathematics and modern physics are testaments to the fact that Kant's monism in that regard was not convincing, even to those most sympathetic to his views in general. Carnap, Lewis and many others freely admitted that there were alternative languages or conceptual schemes. As Rescher put it (1980, p. 337): 'different languages afford us different ways of talking – of saying different sorts of things, rather than saying "the same things" differently or making different claims about "the same thing"'. And in the internal sense, if we are making valid statements about something, then that something must exist, so a straightforward kind of ontological pluralism follows, as discussed in Section 3.4.

Putnam's internal realism was based on the understanding that different groups of humans do routinely develop divergent conceptual schemes. Early on he noted that for the internalist 'there is no God's Eye point of view that we can know or usefully imagine; there are only various points of view of actual persons reflecting various interests and purposes that their descriptions and theories subserve' (Putnam 1981, p. 50). Putnam went as

---

[19] Here I build on the ideas expressed in Chang (2018). Israel Scheffler (1999) has long argued for the compatibility of realism and pluralism. So has Feyerabend, as I argue in Chang (2021).

far as to say: 'internal realism is, at bottom, just the insistence that realism is *not* incompatible with conceptual relativity' (Putnam 1987, p. 17; emphasis original). He gives some instructive examples: 'from the point of view of life and intellectual practice, a theory which treats points as individuals and a theory which treats points as limits may (in their proper contexts) both be right'. Or in physical science, 'a theory which represents the physical interactions between bodies in terms of action at a distance and a physical theory which represents the same situation in terms of fields may both be right'. Generally, 'theories with incompatible ontologies can both be right' (Putnam 1990c, p. 40). Here Putnam begins to sound a bit like the Goodman of *Ways of Worldmaking*: 'That we do not, in practice, actually construct a unique version of the world, but only a vast number of versions ... is something that "realism" hides from us' (ibid., p. 42).

Putnam the internal realist emphasizes that the stability of reference is not a matter of objective truth, but interpretation: 'Why do we regard it as reasonable of Bohr to keep the same word "electron" (*Elektron*) in 1900 and 1934, and thereby treat his two different theories ... as theories which describe the same objects, and regard it as unreasonable to say that phlogiston referred to valence electrons?' (Putnam 1990c, p. 33). And I *do* think it is eminently reasonable to say that phlogiston refers to valence electrons (or conduction electrons)! (Chang 2012a, pp. 43–5). Contrary to the notion of 'rigid designators', Putnam argues: 'reference, like causality, is a flexible, interest-relative notion: what we count as *referring* to something depends on background knowledge and our willingness to be charitable in interpretation. To read a relation so deeply human and so pervasively intentional into the world and to call the resulting metaphysical picture satisfactory ... is absurd' (Putnam 1983, p. 225). Putnam pre-empts the attempt to ground the metaphysical objectivity of reference in the metaphysical objectivity of causation, by noting that causation itself is 'radically perspectival'. The latter is the familiar point about context-dependence: 'For Earthians it may be a discarded cigarette that causes a forest fire, while for Martians it is the presence of oxygen' (Sosa 1993, p. 607). Putnam concludes (1990c, p. 34): 'the claim that we have a notion of reference which is independent of the procedures and practices by which we decide that people in different situations ... do, in fact, refer to the same things ... seems unintelligible'.

Even after renouncing his internal realism, Putnam maintained a commitment to ontological pluralism. In response to Maudlin's push for an 'unsophisticated metaphysical realism', Putnam (2015b, p. 506) stated that he advocated 'sophisticated realism', 'a realism that accepts the idea that the same state of affairs can sometimes admit of descriptions that have, taken at face value, incompatible "ontologies," in the familiar Quinean sense of "ontology".' Similarly, while Niiniluoto (2014, p. 160) does not accept internal realism, his own position of 'critical scientific realism' subscribes to 'the principle of conceptual pluralism', according to which 'all inquiry is relative to some conceptual framework'. In Niiniluoto's view (1999, p. 218), 'the true ingredient of internal realism and the cookie cutter metaphor is conceptual pluralism: the world can be described or conceptualized with several different linguistic frameworks'.

Conservationist Pluralism

If we take a fresh look at the history of science with activist realism in mind, we can see a particularly important way in which epistemic pluralism manifests itself in the practices of working scientists. There is a long-standing unspoken policy among scientists, to which I have given the name of **conservationist pluralism** (Chang 2012a, pp. 218, 224): retain previously successful systems of practice for what they are still good at, and *add* new systems that will give us knowledge about other realities. This practice is quite widespread in actual science, contrary to what is often imagined by standard scientific realists. It is a realistic attitude that practising scientists take. They tend to take care not to discard useful theories from the past – even while declaring them to be false, even while paying lip-service to reductionism and partaking in dreams of the grand unified theory. A whole host of examples of such protected old theories that are put to effective use can be given, ranging from geometric optics to orbital theory in chemistry. And contrary to what one might expect, physics shows this conservationist pattern of development more starkly than any other science. Physicists and others who use physics have retained various successful systems that are good in particular domains: geocentrism (for navigation), Newtonian mechanics (for other terrestrial activities and for space travel within the solar system), ordinary quantum mechanics (for much microphysics and almost all quantum chemistry), as well as special and general relativity, quantum field theory, and more recent theories. For those who would feel upset by the suggestion that the applicability of something like general relativity is 'local', I can only point to all the situations in which we would not dream of using general relativity, as well as the fact that there have been very few specific empirical tests of general relativity. Goodman (1978, p. 4) was correct to observe: 'The pluralist, far from being anti-scientific, accepts the sciences at full value.'

The standard scientific realist inference from success to truth is typically made in a monist framework.[20] According to that reasoning, the most successful of all theories in a given domain is the true theory; when a more successful new theory emerges, the attribution of truth should be withdrawn from the formerly most successful theory. In contrast, realistic realism can make perfect sense of scientists' conservationist behaviour: the success of a theory only tells us that it has a degree of truth-by-operational-coherence; this does not amount to proof, and moreover it does not negate the truth-by-operational-coherence of competing theories. But success does provide a credible promise of continued and further success, a promise to be accepted with our eyes wide open to the problem of induction. That promise of further success can be shared by many competing theories within a given domain. When a system of practice has produced success time and again, it makes sense to keep it for future use. Preserving a successful old system is a modest and reasonable inductivist policy, firmly rooted in the kind of basic inductive reasoning and action that Hume taught us we cannot live without. The successfulness of a tried-and-tested system should be robust in the face of another system that does something else well (or even the same thing well in a different way), which deserves its own credence. To put the point most generally and vaguely: whatever we regard as responsible for success should be preserved so that it may continue giving us that success. With conservationist pluralism we can, once again, understand the progress of science as cumulative: not an accumulation of simple unalterable facts (from which more and more general theories would be formed), but of various locally effective systems of practice which somehow continue to be successful.

To illustrate that point, return to Einstein for a moment, this time to his work on special relativity. Einstein's renunciation of the aether and absolute space and time (see Sections 1.5 and 3.2) should not be taken as a moment of metaphysical enlightenment.[21] Instead, I propose that we take Einstein's work as a pluralist move that demonstrated how such a fundamentally different way of doing physics could be coherent. Most people who have successfully mastered special relativity would remember what a struggle it was to learn to think relativistically, yet how much sense it made once the learning was done. But Einstein did not invalidate the whole range of activities based on the presumption of absolute space and time, which continue to take place in a

---

[20] I have pushed against this monism by recognizing that *success* is a multi-dimensional thing (Chang 2012a, pp. 224–33, and references therein).

[21] Nor should it be seen as the dawning of a general operationalist conscience, a reading which Einstein himself rejected, against Bridgman, Heisenberg and Dingle (see Chang 2009b).

whole range of domains from Newtonian mechanics to molecular biology, not to mention everyday life. Rather, relativity was an amazing move in physics in that it was a piece of maximally unrestricted inquiry, which solved a difficult problem by means of unprecedented methods that led to unforeseen results. It resulted in new pragmatic understanding based in new conceptual activities. The setting-up of the relativistic reference-frame gave new meaning to the very concepts of space and time, and to the concept of simultaneity.

This pluralist perspective has a strong implication for the problem of theory-choice in the philosophy of science. I suggest that the kind of theory-choice (or paradigm-choice) conceived in the traditional way is not necessary. If we stop worrying about choosing *the* winner and eliminating all other competitors, we can have a much more relaxed and open-minded view about the nature and assessment of scientific progress. Pluralists can allow that any system of practice with sufficient promise of progress should be permitted and encouraged. Now, 'sufficient' is a vague notion, of course, and the judgement of sufficient promise depends on how many systems we can afford to maintain at once, within the constraints of the material and cognitive resources available to us. But the vagueness of judgement here is fine, because it suits the vagueness of the actions that need to be taken. Allowing or supporting a system of practice is not a binary decision of life and death; different degrees of support are available, ranging from barely tolerating something to going 'all in' for it. So, if enough people have enough faith that a certain system of practice shows enough promise of progress, then that is good enough prima facie reason for activist realists to consider giving the system some measure of support.

Renouncing monistic theory-choice also means giving up on the notion of progress towards a fixed final destination. Unrestricted inquiry does not have a pre-determined destination because the realities we want to learn about are not pre-figured independently of our inquiry. They become realized in different ways as we push inquiry in different directions. I resist the alluring vision of the development of knowledge steadily con-verging on a final point, which has captivated a range of thinkers from Peirce to Friedman who had insights that would otherwise have led them to pluralism. Instead I have come to embrace a vision of knowledge advanced by Feyerabend, which is a picture of *abundance*. When we let inquiry go unrestricted, its results seem to diverge in interesting ways, even as all inquirers strive to learn in the most successful way possible. We may all build upward on earth, but 'up' is not all the same direction! Going 'up' in Uruguay and going 'up' in Korea are both progressive in the sense of going higher, but they go in precisely opposite directions if we are looking from outer space. The point is not merely that we do not know which direction of development is right, but that there may not even be such a thing as *the* correct or even the best direction of development.

How Different Types of Pluralism Serve Activist Realism

Now I want to consider more carefully how pluralism can serve the cause of activist realism. Here it will be useful to distinguish different types of pluralism, which have different takes on the relation between the different knowledge-systems. ('Knowledge-system' here is a deliberately vague term, to encompass theories, models, research programmes, paradigms and systems of practice.) One important modification I need to introduce is that in addition to the 'tolerant' and 'interactive' forms of pluralism that I previously distinguished, there is another major form: foliated pluralism.

　　According to the weakest type of pluralism, different knowledge-systems that are all valid in the same domain are quite compatible with each other. I adopt Stéphanie Ruphy's (2016) term **'foliated pluralism'** to designate this class of views, which can also be understood as a mild kind of perspectivism (see the 'first layer' of perspectivism that I identified in Section 5.3). According to Ruphy, ontology is enriched by the coherent addition of new perspectives on to existing ones. Foliation provides a contrast to the image of patchwork given by Cartwright: the different layers of knowledge cover the same area, and they are intimately connected with each other, while each adds something different. Does foliated pluralism (or first-layer perspectivism) contribute to activist realism? This very much depends on one's view of the nature of knowledge. If one takes an entirely propositional view of knowledge, one might deny that having two sets of statements that are entirely translatable into each other adds anything to knowledge, compared to the situation in which we have just one of those sets of statements. However, according to my own view, a perspective overlaid upon another one can indeed constitute or create new knowledge, if it supports a new system of practice. But in that case it is quite likely that the different 'leaves' (folios) will start to develop in different directions, eventually creating divergent bodies of active and propositional knowledge. When that happens the neat foliation will cease.

　　In the next type of pluralism, which I have called **tolerant pluralism**, the different knowledge-systems are not presumed to be fully compatible with each other, but they co-exist peacefully without interfering with each other. An excellent example of tolerant pluralism is given in Werner Heisenberg's notion of 'closed theories', as articulated in an insightful presentation by Alisa Bokulich

(2008, ch. 2). According to Heisenberg, classical mechanics and quantum mechanics are both closed theories, meaning that they are each perfect in themselves and cannot be improved in minor ways; they have to be accepted as complete packages, or rejected altogether. In that way closed theories are similar to Kuhnian paradigms, but in Heisenberg's view tolerant pluralism was the correct attitude, rather than the revolutionary abolition of the previous paradigm.[22] Tolerant pluralism can certainly contribute to activist realism. Recall Feyerabend's vision of proliferation. Less flamboyantly, conservationist pluralism points in the same direction: each system of practice can produce its own knowledge, and a tolerant pluralist society can reap the knowledge from all of the systems. However, questions do arise as to how the strands of knowledge emerging from the different systems can be put together. Depending on the situation, the answer may be that they are *not* brought together – different sub-communities may exist in different spheres of activity and not interact with one another in any epistemically meaningful way. Even one and the same person or community may engage in activities in different systems of practice at different times, with no strong connection between them. Teller points out various situations in which physicists use multiple models of the same object and do not try to bring them together: sometimes water needs to be modelled quantum-mechanically, and sometimes as a classical fluid (Teller 2001, pp. 408–9); having both representations involving gravitational force and representations involving space-time curvature gives us 'much richer access to the way things are' (Teller 2018, p. 163).

It is **interactive pluralism** that can serve the needs of activist realism most fully, with productive interactions between the different knowledge-systems generating new avenues of inquiry in addition to those offered separately by each system. I have previously discussed competition, co-optation and integration as three main modes of inter-system interaction (Chang 2012a, sec. 5.2.3). A deeper and richer set of insights about inter-system interactions comes from the close examination of the quantum–classical relation by Bokulich (2008). She sees these interactions as going beyond pluralism, but in my terms her vision is actually in the spirit of the interactive variety of pluralism. Particularly striking is her discussion of 'semi-classical mechanics', which 'uses classical quantities to investigate, calculate, and even explain quantum phenomena', whose methods employ 'an unorthodox blending of quantum and classical ideas, such as a classical trajectory with an associated quantum phase' (Bokulich 2008, p. 104). She stresses that

---

[22] Bokulich (2008, p. 29) notes a similarity to Cartwright's 'metaphysical nomological pluralism'.

there are various types of benefit that have accrued from this line of work, including not only computational convenience but new physical insights and explanations, and even the discovery of new phenomena. Similar kinds of interaction can be seen to be taking place in the GPS example discussed in Section 1.4. Massimi (2018b, p. 356) has also emphasized the importance of productive inter-perspectival interactions: 'Each scientific perspective … functions then both as a context of use (for its own knowledge claims) and as a context of assessments (for evaluating the ongoing performance-adequacy of knowledge claims of other scientific perspectives).' I am not convinced that such interactions can give us non-perspectival knowledge, but they surely create more and better perspectival knowledge. These thoughts extend Sandra Mitchell's (2003; 2020) picture of the ad hoc integration of different systems to meet particular situations. Interactive pluralism is an essential feature of unrestricted inquiry; without it the full potential of our knowledge-seeking activities cannot be realized.

5.5 Epistemic Iteration Revisited

In this section I will consider the iterative character of scientific progress more carefully. The realistic spirit accepts that progress must start from some inherited starting point, and that no starting point given to us will be fully justified. In the process of epistemic iteration, we knowingly start inquiry on the basis of an imperfect starting point, and use the outcome of that inquiry in order to improve its own starting point. The combination of conservatism and optimism in epistemic iteration exemplifies realistic realism. I will start by reviewing and restating my previously articulated ideas about epistemic iteration. And then I will show how epistemic iteration can lead to progressive changes in our concepts and ontologies, our methods and principles, and even our aims. We can recognize all of these patterns of iterative progress with no need for the notion that we get closer to an absolute kind of truth. Accepting a given situation and starting inquiry on that basis is a rational method of achieving progress, which enables deeper kinds of progress than a mere accumulation of facts or a simple increase of precision and scope in our knowledge.

Conservatism and Optimism in Epistemic Iteration

I first articulated the notion of 'epistemic iteration' in relation to the justification of measurement methods (Chang 2004, ch. 5), and have extended the notion in various directions since then (Chang 2007; 2016a; 2017a). I will start by giving a brief and updated review of the idea of progress through epistemic iteration. There are two aspects of epistemic iteration that I want to stress here. First, it is a conservative process in the sense that it accepts *some* inherited state of knowledge as the starting point of our inquiry, recognizing full well that there are probably defects and shortcomings in that starting point. The iterative perspective rejects the familiar foundationalist metaphor of building knowledge on a firm ground. This metaphor is only as good as the flat-earth cosmology which it presumes. But what if we modernized this metaphor? According to our current view of the universe, what we earthlings do is to build outward on a round earth, not upward on a flat earth. The earth is not firmly fixed to anything at all; still, we can build very well on it because it is a large, dense body that attracts other objects. And we build on the earth not because it is the best possible ground to build on in the universe, but because it is where we were born and where we live. That doesn't mean we can't go searching for another planet to build on, but even Elon Musk has to build his spaceships here on earth first. The classical pragmatists were very aware of this situation, and Peirce was in fact one of the original inspirations for my ideas on epistemic iteration. He had a wonderful image for the fact that the actual foundations on the basis of which we conduct inquiry are not entirely firm: inquiry 'is not standing upon the bedrock of fact. It is walking upon a bog, and can only say, this ground seems to hold for the present. Here I will stay till it begins to give way' (Peirce quoted in Misak 2013, p. 34).

The sensible conservatism of epistemic iteration means building on our inheritance and on the achievements of some actual past group of intelligent beings. This basis is something rather than nothing. Even though an initial situation that sets off inquiry is a problematic one, our initial situation embodies some good pre-existing knowledge. This recognition is the basis of what I have called the 'principle of respect' (Chang 2004, p. 43). Inquiry relies on previous knowledge, and then comes back to refine and correct it. This improvement of our starting point is not done on the basis of some absolutely right answer given to us 'from the outside', but on the basis of the best judgement reached as a result of an inquiry process that was reliant on those very assumptions now being corrected. This also means that the validity of the whole inquiry cannot be separated from the basic decency of its starting point. If inquiry ends up negating its starting point *completely*, the whole process is destroyed and we must start a whole new inquiry.

The other aspect of epistemic iteration that I want to stress is a fundamental optimism, that inquiry based on an imperfect starting point will lead to self-improving progress, at least some of the time. I have discussed this process in great detail in the case of temperature and various other concepts from the physical sciences (Chang 2004; 2016a), but an initial illustration can be made with a simple example-and-metaphor taken from everyday life. Without wearing my glasses, I cannot see small things very well. So, if I pick up my glasses to examine them, I am unable to see the fine scratches on the lenses. But if I *put on* those same glasses and look at myself in the mirror, I can see the details of the lenses quite well. In short, my glasses can show me their own defects, giving me a more refined view than I can obtain without using them. But how can I trust the image of defective glasses that is obtained through the very same defective glasses? In the first instance, my confidence comes from the *apparent* clarity and acuity of the image itself, regardless of how it was obtained (as I can also achieve by squinting, looking at a distant object). That gives me some reason to accept, provisionally, that the defects I see in my glasses *somehow* do not affect the quality of the image (even when the image is of those defects themselves). But sometimes the defects will of course affect what I see. There may be a black dot in the middle of the left lens, or a smudge in the middle of both lenses, and in such cases I would know how all the images I have with the glasses on should be modified, including how the glasses themselves look.

Peirce (1934, pp. 399–400) saw that good inquiry was a process of self-correction, drawing his intuition on that score from *mathematical iteration*, a procedure in which one knowingly starts with an answer that is not quite right, and then applies an algorithm that uses the imperfect answer in order to arrive at a better one. In empirical inquiry, too, when we are fortunate enough, we are able to enter into a process of successive approximations to a convergent outcome. I have demonstrated this kind of possibility, especially with the example of the measurement of Kelvin's absolute temperature (Chang 2004, ch. 4; Chang and Yi 2005). Peirce's insights pointed to an inherently progressivist yet non-foundationalist epistemology. Dewey, too, had an iterative conception of the 'self-corrective process of inquiry', focusing especially on the development of methods: 'The problem . . . is whether inquiry can develop in its own ongoing course the logical standards and forms to which *further* inquiry shall submit. One might reply by saying that it *can* because it has' (Dewey 1938, p. 5; emphases original).

Conceptual and Ontological Change by Iteration

Although I originally developed the idea of epistemic iteration in relation to the justification and improvement of measurement methods, I soon came to see its broader relevance. An immediate recognition was that changes in methods and standards of measurement affected the very meaning of the concept involved: the development of the concept of temperature was closely tied to the development of thermometry. Generally speaking, the very concepts with which we think become refined and corrected through our practical scientific engagement with nature by means of them, undergoing subtle and sometimes not-so-subtle changes in meaning. And concepts are directly linked to ontology, since realities are the referents of concepts that facilitate coherent activities. As a result of successful inquiry we can arrive at a better ontology, a better set of realities to think about and live with. Some careful reflections are needed on how exactly we do identify and stabilize real entities. It will be useful to discuss some cases from chemistry and physics, because those sciences are often taken to supply the best examples of timeless natural kinds that are characterized in an essentialist manner.

For example, the concept of 'atom' in physics and chemistry went through very fundamental changes. Initially the notion of 'atom' meant a fundamental unit of matter that was unalterable and indivisible. This ancient concept was successfully adapted to make the foundation of the atomic–molecular theory, in which each chemical element was presumed to be made up of a distinct type of atom, and each compound substance a unique combination of mixed species of atoms. On this basis a very successful system of chemistry developed, with each of the many thousands of distinct chemical substances being assigned a definite molecular formula. However, this system highlighted a new question, which it had no conceptual resource to answer: why (and how possibly) did atoms stick together to form a molecule? In the end this problem was only solved by postulating an internal structure to atoms, in particular allowing subatomic electrons to be shifted around or shared between neighbouring atoms. So the indivisible atom that was the foundation of atomic–molecular chemistry matured into something fundamentally different, maintaining its identity when it is by itself but becoming altered when it enters into combination.

Our dealings with particular substances also proceed by epistemic iteration. For a very mundane case take the concept of 'fixed air', or carbon dioxide in modern terms (see Chang 2016a, sec. 3 for further details). This is a substance that has always been with humanity (in our exhaled breath, and more recently in bubbles in beer) but not identified clearly as a chemical substance until the second half of the eighteenth century. The concept of 'fixed air' initially arose from the work of Stephen Hales, who showed that gases ('airs') could be 'fixed' in solid substances; any gas released from such solid combinations was called 'fixed air' (though I think it should really have been called 'unfixed air' – air *released* from a previous stated of fixedness). It was Joseph Black's pioneering work that started the process of turning 'fixed air' into a more precise category. As he studied the chemical reactions of fixed air produced from chalk, Black noted that it produced a white precipitate when it came into contact with lime-water (solution of lime — the mineral, not the fruit). This was a very convenient reaction with a clear and fail-safe result, which came to serve as a good operational definition of fixed air.[23] Next he applied the lime-water test to various other gases, and learned that in fact not all gases released in chemical reactions passed the lime-water test, though they were all 'fixed air' according to the original meaning of the term.[24] What Black did here may seem mundane, but it was a very serious revision of the concept of 'fixed air'. He started with the old notion of fixed air as 'air released through chemical reactions from solid substances', collected and studied such gases, and recognized that *some* of those gases had a peculiar and interesting behaviour. Adopting that behaviour as definitive of fixed air meant a considerable narrowing of the concept, and created the impetus to identify other more-specific gases (according to their own characteristic behaviours). Black made progress by latching on to an island of unnamed regularity (*some stuff* here precipitates lime-water) and using that regularity in order to make a new stipulation of the meaning of 'fixed air'. After this step, further and more precise properties of fixed air and other distinct gases could be determined precisely, leading to an improved stage of knowledge. The more narrowly defined fixed air went on to be recognized as an oxide of carbon, and then more specifically as carbon dioxide (CO_2).

On reflection it is clear that epistemic iteration is a plausible method of making conceptual (and ontological) development. To launch any kind of inquiry we have to have some sort of concepts to begin with; otherwise no questions can be formulated, no reasonings made, no experiments designed and no conclusions drawn. Basic concepts are not something we can deny ourselves until science has sufficiently progressed. In the here and now of science, we need to work with operational realities designated by concepts that we already have, so we start with the best concepts that we currently

---

[23] In modern terms, lime-water is an aqueous solution of calcium hydroxide ('slaked lime'), which reacts with carbon dioxide to produce calcium carbonate (chalk), which is insoluble in water so makes a precipitate. In chemical symbols, the reaction is: $Ca(OH)_2 + CO_2 \rightarrow CaCO_3 + H_2O$. It was known from early on that chalk ($CaCO_3$) could be turned into caustic lime (CaO) by heating, but it was not recognized until Black's work that the process was a decomposition of chalk into lime and fixed air (CO_2). Caustic lime (CaO) becomes slaked lime ($Ca(OH)_2$) by absorbing water (H_2O), and slaked lime dissolves in water, making lime-water. See Lowry (1936, p. 61) and the rest of ch. 4.

[24] It is telling that Lavoisier had originally expected to generate fixed air in the operation that produced oxygen (heating a metallic calx). Without the lime-water test he might have believed that he had produced fixed air; knowing that this gas was not fixed air was a key to the rethinking that led to his identification of oxygen.

possess, and get on with the inquiry. The results of our investigation can make us rethink the concepts we originally began with. Any considerable and lasting success in our activities generates confidence in the concepts used in those activities, and invites us to consider them as 'natural', inspiring the talk of natural kinds. The key here is to recognize entities designated by such good concepts as features of the world-we-live-in, without falling into the fallacy of pre-figuration (see Section 2.1). And what is considered natural will evolve along with the general progress of science.

Progress in Methods and Principles

Iterative progress can also result in the improvement of methods and principles of inquiry. This applies to all manner of rules that are used in science – including methods of measurement, hypothesis-testing, mathematical analysis, theory-construction, modelling and simulation, and logical reasoning. In fact this is not an entirely separate matter from conceptual change, since a concept is essentially bound up with the rules for its use. This is very clear when a concept has an explicit definition, but any concept in coherent use is bound by some recognizable rules of use, including both semantic and methodological rules. With updated rule-bound concepts we can pose different and more instructive questions about the same subject matter. And since inquiry will create new meanings, there is no guarantee that the validity of well-known principles will not be affected as a result of inquiry (recall the example of 'An atom is indivisible', just discussed above).

In the process of iterative inquiry even the most fundamental rules, including those regarded as a priori principles, are subject to modification depending on how inquiries based on them work out. As Stump (2015) discusses in insightful detail, the actual variability of purportedly a priori principles has been recognized by a diverse array of thinkers over many decades. Predating the scope of Stump's survey, Whewell recognized that over the course of the history of science the 'fundamental ideas' and the principles associated with them had changed and evolved (see Losee 1993, pp. 126–34). In recent years the most prominent philosopher articulating a similar view has been Friedman (2001). Dewey stressed that there were no absolute standards governing inquiry and that all methodological rules, including even logical principles, were only 'operationally a priori with respect to further inquiry' (Dewey 1938, p. 14; emphasis original). C. I. Lewis broadly agreed with Dewey's view of logic, and both thinkers extended the same perspective to cover all a priori principles (see Sections 1.6 and 3.2).

From the perspective of epistemic iteration, 'a priori' just means being *prior*, in the sense of being accepted without question at the start of each inquiry, rather than prior to all experience – and who among us can tell what anything is like prior to *all* experience? This is intended to 'lower the tone' in a helpful way (like a drop-shot in tennis helpfully slowing things down), which also allows us to escape the complex historical baggage of the 'a priori' term. Here is Elgin (2017, p. 64): 'Inquiry, as Quine insists, always begins *in medias res*. We start with opinions, values, methods, and standards that we consider relevant and that we are inclined to credit. Although we recognize that they are less than wholly satisfactory as they stand, they comprise our current best take on the matter under investigation.'

It is important to note that rules are not just changeable, but capable of self-improvement in the standard manner of epistemic iteration. We start a process of inquiry on the basis of some prior principles. And then the result of the inquiry may end up suggesting changes in the principles, and the updated principles will be part of the coherentist reflective equilibrium that is Elgin's version of pragmatist validation (2017, ch. 4). C. I. Lewis was very clear on this point: the justification of the choice of any conceptual system can only accrue from the experience of trying to apply the system in question to various areas of inquiry (1929, pp. x–xi). Even though Lewis says that the a priori is legislated by us (and therefore analytic, by choice), he makes it very clear that legislation should have good reasons behind it. The good reasons come down to operational coherence. If we are being rational, our choice of activities will eventually be determined by what works out. Lewis states (1929, p. 239, emphases added): 'while the a priori is *dictated* neither by what is presented in experience nor by any transcendent and eternal factor of human nature, it still *answers to* criteria of the general type which may be termed pragmatic'. So it goes in inquiry: we start with some ground rules, but according to what we learn through working under those rules, we can make informed methodo-logical decisions about how those rules ought to be adapted. It is an important part of the thorough empiricism of pragmatism (see Section 1.6) to treat methodology like anything else that we learn empirically.

The Iterative Development of Aims

Not only our concepts and methods but our aims, too, can evolve through the process of epistemic iteration. Activities can result in the alteration of their own aims, and the epistemic activities involved in inquiry are no exception in that regard. Agents are not constrained to continue pursuing the same aims, especially in the long run. The relevant point in relation to epistemic iteration is that it makes sense to stop trying to achieve an aim that turns out not to be feasible; it is rational to adjust our aims in a realistic spirit. In the other direction, something that we can do coherently and sustainably often becomes something we *want* to be doing. Revised aims will in turn reshape the activities intended for achieving them, whose outcomes may lead to further revisions in the aims we want to pursue. This iterative co-evolution of activities and aims can continue indefinitely.

An iterative process is necessary for the development of aims because we cannot judge whether an aim can be plausibly pursued, without having tried to achieve it. This is a parallel point to the lack of an indubitable foundation in the justification of knowledge. So we need to try constructing activities designed to achieve an aim that initially seems achievable and desirable, and be guided by the outcomes. But does the abandonment of an aim in itself constitute progress? It can, in the sense that pursuing an unachievable aim is not a productive use of our efforts. And it is not the case that the only thing we can do with an implausible aim is to abandon it altogether. Often the outcome of inquiry is the adjustment of the aim so that it becomes more plausible and productive. Even the renunciation of an aim can be turned into a positive basis for further developments. For example, in the early days of science it seemed to many people that making a perpetual-motion machine was a plausible aim. And who would have had a firm basis on which to pronounce it impossible? Numerous attempts were made, and many things were learned in the course of this activity. In the end the aim was abandoned, but what happened was not a simple admission of defeat. Instead, the impossibility of perpetual motion was now turned into the theoretical foundations of the new science of thermodynamics, in the form of the first and second laws of thermodynamics. The failure of an initial aim can be incredibly important in this way, leading to a great deal of scientific progress. The same lesson can be read in Einstein's abandonment of the aim of explaining the failure to detect the motion of the earth through the aether.

Generally we can observe a useful rational interaction between the success of scientific activities aimed at establishing truths about unobservable entities, and scientists' and philosophers' confidence in pursuing that aim.[25] For example, I think it was quite reasonable for Poincaré, Duhem and Mach to be wary of the realist aims of microphysics around the end of the nineteenth century, having seen the failure of attempts to engage directly with atoms. And it was also reasonable for many other scientists to lose such restraint with

---

[25] For thoughtful discussions about the rationality in the context of pursuit, see Šešelja, Kosolosky and Straßer (2012), and Šešelja and Straßer (2014).

the arrival of more direct means of engaging with atomic and subatomic particles in the early twentieth century. We can detect some healthy responses to the evolving scientific situation in philosophers' debates, too, with some time lag. I think the rise of scientific realism in the middle of the twentieth century was a reasonable but naïve response to the successes of microphysics, molecular biology and other sciences. More mature was van Fraassen's (1980) constructive empiricism, which took a long view on the development of science and suggested moderating the youthful ambition of science by distinguishing truth and empirical adequacy, taking only the latter as a plausible aim in relation to unobservables. But other commentators have proposed insightful modifications to van Fraassen's pessimism, identifying circumstances in modern science under which we can be optimistic about learning *some* truths about *some* unobservable entities. Hacking's (1983) experimental realism is a good example of this modulation, consonant with the intuitions of practising scientists.

5.6 Progress and the Scientific Realism Debate

Looking at the scientific realism debate from the standpoint of activist realism, the key consideration is whether and how well various philosophical positions promote scientific progress. This perspective makes an interesting realignment of positions. Some of the positions traditionally considered anti-realist can be quite conducive to the promotion of progress, so they should be considered realist by my lights. On the other hand, some aspects of standard scientific realism are actually unhelpful to scientific progress, or they promote an impoverished version of scientific progress, so they should be considered anti-realist. What I am proposing here is not just a playful relabelling of positions, but a reframing of the traditional debates that will help us keep in mind why it matters to be 'realist'.

How 'Anti-Realism' May Align with Activist Realism

One key concern in this chapter has been the re-examination of the notion of scientific progress, which I take very generally at the outset as the enhancement of scientific knowledge. As indicated in Section 5.1, the shape of scientific progress is much more abundant than usually imagined by philosophers. In Sections 5.2–5.5 I have made further observations on the character of scientific progress seen from the activist realist standpoint – internal/perspectival, iterative and pluralistic. Armed with this view of scientific progress, I now want to return to the debate concerning scientific realism, and examine how the different positions stand in relation to scientific progress. This will help me further differentiate activist realism, which is unequivocally committed to progress, from some other positions that are not so committed. And the differential implications for the promotion of progress indicate that the scientific realism debate has not all been an empty dispute about words and slogans. The realignment of different positions that will emerge in this light is going to have some surprises in store.

To start, let's admit that anti-realism often does produce injunctions against the pursuit of certain types of knowledge that can actually be pursued perfectly cogently. This is reminiscent of the legend that on the Pillars of Hercules (the promontories that flank the Strait of Gibraltar) there was the inscription 'Non plus ultra' (do not go beyond), keeping ships confined to the Mediterranean away from the unknown hazards of the Atlantic. In anti-realist views on science there are often similar injunctions, which amount to various ways of blocking the road of inquiry. For example, positivism had a tendency to refuse to theorize at all about unobservables, as in the famous case of Ernst Mach's objection to atomism. As a general policy such anti-realism does place a needless restriction on research activity. Theories about unobservable entities do usually have observational consequences, so the refusal to theorize about them results in a loss of opportunities for empirical learning.

On the other hand, it is not the case that all philosophical positions that are normally considered anti-realist serve to hinder the generation of new knowledge. Empiricist positions are often regarded as anti-realist because they do not commit to the existence of unobservable entities. However, the empiricist insistence on making concepts meaningful by reference to observations has always been motivated by the desire to ensure that our statements have empirical content (and therefore can contain actual knowledge). In that sense, empiricist doctrines ought to be regarded as species of activist realism. In the early twentieth century, operationalism and verificationism sought to guard against rationalist tendencies in science and philosophy that led to the loss of empirical content, which works against activist realism. And it is, of course, not the case that theoretical progress always comes from resorting to unobservables. A good deal of creative and productive scientific research has been done by focusing on the observable, for example by Mach on acoustics, by Joseph Fourier on heat conduction (producing in the process the method of Fourier analysis), and by Sadi Carnot on heat engines (laying down the foundations of all subsequent theories of thermodynamics).

One interesting illustration of the subtlety involved in discerning what is realist in the activist sense is the debate between van Fraassen and

Hacking, mentioned at the end of Section 5.5. At the surface level I am quite clearly with Hacking: knowledge can credibly be extended to the unobservable realm, and it is the job of science to do that, rather than limiting its truth-seeking to the observable realm. Trying to pronounce a priori where knowledge can and can't be extended is unwise, and contrary to the imperative of progress. 'Unobservables can't be learned about' is just not a plausible general empirical prediction concerning the future of science. The entire history of modern science is a testament to people's ability to learn about what is unobservable in van Fraassen's sense, including X-rays, viruses, genes, nano-particles and whatnot. However, at a deeper level I am also sympathetic to van Fraassen, and I do not think that the difference between their positions is as stark as it might at first appear. What van Fraassen calls 'empirical adequacy' is not so different from what I call truth-by-operational-coherence, which could be seen as a pragmatist version of empirical adequacy. So van Fraassen's position could be repurposed as a sort of pragmatist realism in my sense, with no hindrance to progress. It is also worth noting that Hacking's entity realism does not deliver the kind of reality that metaphysical realists want; as I have discussed in Chapter 3, Hacking's notion of reality is actually very close to my notion of reality based on operational coherence.

How Standard Scientific Realism May Not Promote Progress

Turning to standard scientific realism now (see Section 2.1 again for its characterization), I want to sound some notes of caution about how well it can be aligned with activist realism. One difficulty is that the state of scientific knowledge envisaged in standard scientific realism is highly unrealistic. Not only can it be fairly disputed that our actual science has attained the kind of knowledge that standard scientific realists attribute to it, but it is also unreasonable to demand that scientists should pursue it. Being unrealistic gets in the way of the activist dimension of realism. In response to this difficulty, a number of philosophers in recent decades have tried to maintain standard scientific realism by weakening it sufficiently to escape common anti-realist arguments. The resulting positions, which I think of (uncharitably) as watered-down standard scientific realism, may seem congenial to my own position because they make standard scientific realism more realistic. However, I do not advocate such positions because they are not sufficiently progressivist. They are focused on presenting *existing* scientific knowledge in the most favourable light, and this focus ends up serving as a distraction that interferes with the active promotion of progress.

Take the common watering-down move of claiming that our best scientific theories are only *approximately* true. Even though I fully share the desire to move away from black-and-white judgements of truth and falsity, I do not wish to go along with saying that what our scientific theories achieve is an approximation to *correspondence-truth*. This is because correspondence-truth as standard scientific realists conceive it is not an operational notion (see Chapter 2). Let us be clear: we can see that Galileo's law of free fall is an approximation to the full Newtonian solution to the problem of falling bodies near the surface of the earth, but that says nothing about whether either Galileo or Newton gives us something that approximates the ultimate Truth. Similarly with how the equations of Newtonian mechanics approximate those of special relativity. Approximation, as an actually verifiable process of getting closer to a known goal, would be a progressive act. But Galileo did not craft his law with the aim of getting it to come out close to the verdict of Newtonian mechanics. The idea of approximate truth so prized by the standard scientific realists is only a retrospective rationalization, not a driver of scientific progress. In contrast, my notion of truth-by-operational-coherence is an actually pursuable aim, which can be increased by degrees. It may be improved as an outcome of further inquiry, and its improvement can actually be tracked. There is no need to deal with anything equivalent to the inaccessibility of Truth about transcendent Reality, and no need to hedge the attribution of truth with the false talk of 'approximation'.

Let me now return to the core standard scientific realist argument from the success of science, which is usually put in the form of an inference from the success of a theory to its truth. This is seen as an 'ampliative' inference, since the conclusion contains more content than the premise. A simple way to understand what is going on here is that some standard scientific realists have turned realism into a game of plausibly claiming to know more than we actually do. This is a difficult game to win at, so there is plenty of challenging work to do. However, in practice, 'ampliative inference' is a lazy notion: if only, somehow, just by thinking, we can get stronger conclusions than apparently warranted by the evidence we put into the thinking ...

Ampliative modes of inference, which include induction and abduction, do have important roles to play in the actual processes of empirical inquiry. But it should be recognized that they are *inferences* not in the sense of logical deductions, but more in the sense of everyday usage, meaning 'a guess that you make or an opinion that you form based on the information that you have' (*Cambridge Dictionary*, online). Such inferences do not in themselves produce knowledge. Rather, they guide inquiry by pointing us in the direction of possible future knowledge. As Igor Douven notes, abduction in

Peirce's original conception was meant as a method of discovery (hypothesis-generation), though in current discussions it is usually taken as a method of justification (hypothesis-testing).[26] I think we would do well to return to Peirce here. I propose that we take realism as an enterprise of **ampliative inquiry** in which knowledge is enhanced through hard empirical work. This is something entirely different from the business of armchair ampliative inference (conducted by philosophers or scientists). Activist realism is not about inferring truth from success, but about pursuing success in meeting various aims and being led to various truths in the course of that effort. The role of philosophers should be to encourage ampliative inquiry done by scientists, and even to engage in it ourselves where the scientists do not, in the spirit of **complementary science** as I have called it (Chang 2004, ch. 5; 2012b).

---

[26] See Douven (2021), especially the supplement, 'Peirce on Abduction'. He quotes Peirce as follows: 'Abduction is the process of forming explanatory hypotheses. It is the only logical operation which introduces any new idea' (Peirce 1934, §5.172).

Closing Remarks

A Humanist Vision of Knowledge

This book has been an attempt to inject more pragmatism into the philosophy of science, and into epistemology and metaphysics especially as they bear on the philosophy of science. From a pragmatist standpoint I have proposed renewed conceptions of knowledge, truth and reality as notions that are fully meaningful in practice. I have also wanted to put pragmatism into a broader framework, namely that of humanism. When applied to the philosophy of science, humanism means treating science as a human enterprise, undertaken by people in order to meet people's needs – intellectual, material and social. As explained in Section 1.6, by 'humanism' I do not mean human chauvinism. For non-human agents there will be different pragmatist epistemologies, which are bound to differ significantly from human pragmatist epistemology. In my view, that only enhances the need for us to understand the nature of our own human ways of knowing, before and while we attempt to understand other kinds of knowing. This humanist framing shapes my interpretation of pragmatism, which focuses on two main aspects.

The first focus is on empiricism, which I take to be the stance that ultimately the only source of learning available to us humans is human experience, and that in matters of knowledge experience should have higher authority than any other alleged arbiter. Empiricism is the fundamental credo of humanist epistemology; some sort of empiricism is the inevitable starting-point of epistemology in our secular humanist age, as much as the presumption of God would have been the inevitable bedrock of any intellectual discourse in an earlier age in Europe. Pragmatism rejects, strenuously and consistently, other sources and criteria of knowledge. Such relentless empiricism should be well-rounded, too, taking a full view of experience as lived experience. Empiricism is significantly perverted when it is taken to imply that we should assign epistemic authority only to direct results of sense-perception.

The second focus in my interpretation of pragmatism is a reminder of something obvious: pragmatism is an action-oriented philosophy. There are two dimensions to the action-orientation. First, all pragmatists would agree that we should pay attention to practical consequences of knowledge. More fundamentally, we need to understand the nature of knowledge itself, and inquiry itself, in the domain of action. We need a pragmatist elucidation of epistemic practices in terms of actions: what do we actually *do* in order to gain knowledge, to test it and to improve it, and to use it? How best can we organize and support such epistemic actions that we engage in? If we conceive pragmatism generally as a philosophical commitment to engage with practices, then pragmatist epistemology will concern itself with all practices relating to knowledge.

Empiricism and action-orientation in pragmatism are deeply connected with each other. This is because lived experience takes place only in the context of action. There is no such thing as truly passive observation. Pragmatism considers the nature of doings in our lives: why we do them, how it is possible to do them, and how we can do them better. Knowledge is knowing how to do things. When pragmatism is applied to the philosophy of science, it turns the latter into the philosophy of scientific practice, considering the nature of the doings that constitute scientific work, and the purposes and consequences of these doings. Work requires the virtue of humility, accompanied by social cooperation and toleration. When we take a humble pragmatist view of our human situation, pluralism results naturally. Our work in attaining and improving knowledge will go in many directions and occur in various systems of practice, each of them with only partially confirmed potential to produce good knowledge. We make progress by our efforts to develop our chosen systems, in competition and in collaboration with each other. That is how we get on in human life in general, and science is no exception.

The Road Ahead

When I first went to study in the United States, I was struck by the fact that in many American schools and universities the graduation ceremony was called 'commencement': you've completed your course of study, and the end of that is now the beginning of the next stage. So it should be with a book, too. I have no grand *conclusions* to pronounce, and I also cannot add at this point any significant synthesis of what I have discussed so far, since the main body of the book has continually presented an evolving synthesis. Instead, what I want to do now is to indicate some important tasks lying ahead, in two broad categories. First, there are many issues I wish I could have dealt with more convincingly and knowledgeably. The following is a small list of areas of future work for myself, for interested colleagues, and for talented students.

- What, really, is *practice*?
- What is the nature of the *pragmatic understanding*, which is the foundation of operational coherence?
- How can *phenomenology*, and the philosophy of *cognition*, help with a really full understanding of scientific observation and experiment?
- How do we advance an improved theory of *action* fully incorporating values and judgements?
- How can my pragmatist philosophy of science be connected more convincingly with *social epistemology* and the sociology of knowledge?
- How best can we connect the assessment and improvement of *values and aims* more closely with the epistemological and metaphysical discussions given here?

Secondly, there are some obvious ways in which I hope the ideas and perspectives presented in this book can be put to good use. My ideas of reality and truth may be applied in order to understand and improve existing discourses in various areas. I also hope that my ideas will facilitate the engagement of philosophy with other scholarly disciplines and the various practices of life. More critically, my ideas are intended to serve the function of discouraging philosophical debates based on notions that do not connect with practices of life, and channelling the energy and talents of philosophers more decisively into productive directions.

Bringing Philosophy Back to Life

In its content, format and tone, this book is a sincere expression of how I see the role of philosophy in human life. A century ago Dewey called for a 'recovery of philosophy', and with this book I hope to have made my own modest contribution to that project. His main motivation is worth quoting at some length:

> But what serious-minded [people] not engaged in the professional business of philosophy most want to know is what modifications and abandonments of intellectual inheritance are required by the newer industrial, political, and

scientific movements . . . Unless professional philosophy can mobilize itself sufficiently to assist in this clarification and redirection of [people's] thoughts, it is likely to get more and more sidetracked from the main currents of contemporary life. This essay may, then, be looked upon as an attempt to forward the emancipation of philosophy from too intimate and exclusive attachment to traditional problems. It is not in intent a criticism of various solutions that have been offered, but raises a question as *to the genuineness, under the present conditions of science and social life, of the problems.* (Dewey 1917, p. 5; emphasis original)

A meaningful kind of philosophy must regard intelligence 'in consideration of a desirable future and in search for the means of bringing it progressively into existence' (ibid., p. 29). Pragmatism should be a philosophy that helps us to do things better in life, not just to think more clearly about what words mean. In fact, semantics itself should be recognized as a tool for effective action.

In this book I have expressed my desire for a philosophy of science that actually promotes the progress of scientific knowledge, and I have made my best effort to realize that ideal. I have adopted pragmatism as the best general framework for such a philosophy of science. Adopting pragmatism should not mean abandoning the pursuit of truth and reality. On the contrary, it means revising the philosophical concepts of truth and reality so that they become meaningful and achievable in practice. The pursuit of such truths, the pursuit of knowledge about such realities, is what I have proposed here as 'realism for realistic people'.

Bibliography

Ahmad, Zubair, Rahim, Shabina, Zubair, Maha and Abdul-Ghafar, Jamshid. 2021. Artificial Intelligence (AI) in Medicine, Current Applications and Future Role with Special Emphasis on Its Potential and Promise in Pathology. *Diagnostic Pathology* 16, article 24.

Ankeny, Rachel and Leonelli, Sabina. 2016. Repertoires: A Post-Kuhnian Perspective on Scientific Change and Collaborative Research. *Studies in History and Philosophy of Science A*60: 18–28.

Arabatzis, Theodore. 2006. *Representing Electrons: A Biographical Approach to Theoretical Entities*. University of Chicago Press.

Audi, Robert. 2014. *Epistemology: A Contemporary Introduction to the Theory of Knowledge*. London: Routledge.

Austin, J. L. [1950] 1979. Truth. In J. O. Urmson and G. J. Warnock (eds.), *Philosophical Papers*, 3rd edn, 117–33. Oxford University Press.

[1957] 1979. A Plea for Excuses. In J. O. Urmson and G. J. Warnock (eds.), *Philosophical Papers*, 3rd edn, 175–204. Oxford University Press.

1962. *How to Do Things with Words*. Oxford: Clarendon Press.

Auxier, Randall E., Anderson, Douglas R. and Hahn, Lewis Edwin (eds.). *The Philosophy of Hilary Putnam*. Chicago and La Salle, IL: Open Court.

Baghramian, Maria (ed.). 2012. *Reading Putnam*. New York: Routledge.

Barnes, Barry. 2000. *Understanding Agency*. London: Sage.

Beck, Lewis White. 1968. The Kantianism of Lewis. In Paul Arthur Schilpp (ed.), *The Philosophy of C. I. Lewis*, 271–85. La Salle, IL: Open Court.

Bergson, Henri. [1896] 1912. *Matter and Memory*. London: Macmillan.

[1907] 1911. *Creative Evolution*. New York: Henry Holt.

Berkeley, George. [1709] 1910. *A New Theory of Vision and Other Writings*. London: J. M. Dent & Sons.

Bermejo, Fernando, Hüg, Mercedes X. and Di Paolo, Ezequiel A. 2020. Rediscovering Richard Held: Activity and Passivity in Perceptual Learning. *Frontiers in Psychology*, published online 19 May 2020. https://doi.org/10.3389/fpsyg.2020.00844

Bokulich, Alisa. 2008. *Reexamining the Quantum-Classical Relation: Beyond Reductionism and Pluralism*. Cambridge University Press.

Boon, Mieke. 2015. Contingency and Inevitability in Science: Instruments, Interfaces, and the Independent World. In Léna Soler *et al.* (eds.), *Science as It Could Have Been: Discussing the Contingency/Inevitability Problem*, 151–74. University of Pittsburgh Press.

Boswell, James. 1935. *Boswell's Life of Johnson*, 6 vols. Oxford University Press.

Bowdle, Brian F. and Gentner, Dedre. 2005. The Career of Metaphor. *Psychological Review* 112: 193–216.

Boyd, Richard. 1990. Realism, Approximate Truth, and Philosophical Method. In C. Wade Savage (ed.), *Scientific Theories*, 355–91. Minneapolis: University of Minnesota Press.

1999. Homeostasis, Species, and Higher Taxa. In Robert A. Wilson (ed.), *Species: New Interdisciplinary Essays*, 141–86. Cambridge, MA: MIT Press.

Brading, Katherine and Crull, Elise. 2017. Epistemic Structural Realism and Poincaré's Philosophy of Science. *HOPOS* 7: 108–29.

Brading, Katherine and Landry, Elaine. 2006. Scientific Structuralism: Presentation and Representation. *Philosophy of Science* 73:571–81.

Bradley, Richard. 2017. *Decision Theory with a Human Face*. Cambridge University Press.

Brandom, Robert. 1994. *Making It Explicit*. Cambridge, MA: Harvard University Press.

Bridgman, Percy Williams. 1927. *The Logic of Modern Physics*. New York: Macmillan.

1940. Science: Public or Private? *Philosophy of Science* 7/1: 36–48.

1955. *Reflections of a Physicist*. 2nd edn. New York: Philosophical Library.

1956. The Present State of Operationalism. In Philipp Frank (ed.), *The Validation of Scientific Theories*, 75–83. Boston, MA: Beacon Press.

1959. *The Way Things Are*. Cambridge, MA: Harvard University Press.

Brown, Matthew J. 2012. John Dewey's Logic of Science. *HOPOS* 2: 258–306.

2013. Values in Science beyond Underdetermination and Inductive Risk. *Philosophy of Science* 80: 829–39.

2020. *Science and Moral Imagination*. University of Pittsburgh Press.

Brun, Georg. 2016. Explication as a Method of Conceptual Re-engineering. *Erkenntnis* 81:1211–41.

Buber, Martin. [1923] 1937. *I and Thou*. Edinburgh: T. & T. Clark.

Burge, Tyler. 1998. Computer Proof, Apriori Knowledge, and Other Minds: The Sixth Philosophical Perspectives Lecture. *Philosophical Perspectives* 12: 1–37.

Button, Tim. 2013. *The Limits of Realism*. Oxford University Press.

Cao, Tian Yu and Schweber, Silvan S. 1993. The Conceptual Foundations and the Philosophical Aspects of Renormalization Theory. *Synthese* 97: 33–108.

Cappelen, Herman. 2018. *Fixing Language: An Essay on Conceptual Engineering*. Oxford University Press.

Capps, John. 2019. The Pragmatic Theory of Truth. In *Stanford Encyclopedia of Philosophy*. https://plato.stanford.edu/archives/sum2019/entries/truth-pragmatic.

Carnap, Rudolf. 1950. Empiricism, Semantics, and Ontology. *Revue Internationale de Philosophie* 4/11: 20–40.

1963. Replies and Systematic Expositions. In Paul Arthur Schilpp (ed.), *The Philosophy of Rudolf Carnap*, 859–1013. La Salle, IL: Open Court.

Carrier, Martin. 2013. Values and Objectivity in Science: Value-Ladenness, Pluralism and the Epistemic Attitude. *Science & Education* 22: 2547–68.

Cartwright, Nancy. 1983. *How the Laws of Physics Lie*. Oxford: Clarendon Press.

1999. *The Dappled World: A Study of the Boundaries of Science*. Cambridge University Press.

2011. A Philosopher's View of the Long Road from RCTs to Effectiveness. *The Lancet* 377 (9775): 1400–1.

2019. *Nature, the Artful Modeler: Lectures on Laws, Science, How Nature Arranges the World and How We Can Arrange It Better*. Chicago: Open Court.

Cartwright, Nancy and Hardie, Jeremy. 2012. *Evidence-Based Policy: A Practical Guide to Doing It Better*. Oxford University Press.

Cassirer, Ernst. [1910] 1953. *Substance and Function & Einstein's Theory of Relativity*. New York: Dover.

Chakravartty, Anjan. 2010. Perspectivism, Inconsistent Models, and Contrastive Explanation. *Studies in History and Philosophy of Science* 41: 405–12.

2017. *Scientific Ontology: Integrating Naturalized Metaphysics and Voluntarist Epistemology*. New York: Oxford University Press.

Chalmers, Alan. 2009. *The Scientist's Atom and the Philosopher's Stone*. Dordrecht: Springer.

2013. *What Is This Thing Called Science?* 4th edn. St Lucia: University of Queensland Press.

Chalmers, David J. 1996. *The Conscious Mind: In Search of a Fundamental Theory*. New York: Oxford University Press.

Chang, Hasok. 2002. Rumford and the Reflection of Radiant Cold: Historical Reflections and Metaphysical Reflexes. *Physics in Perspective* 4: 127–69.

2003. Preservative Realism and Its Discontents: Revisiting Caloric. *Philosophy of Science* 70: 902–12.

2004. *Inventing Temperature: Measurement and Scientific Progress*. New York: Oxford University Press.

2007. Scientific Progress: Beyond Foundationalism and Coherentism. In Anthony O'Hear (ed.), *Philosophy of Science*, 1–20. Cambridge University Press.

2008. Contingent Transcendental Arguments for Metaphysical Principles. In Michela Massimi (ed.), *Kant and the Philosophy of Science Today*, 113–33. Cambridge University Press.

2009a. Ontological Principles and the Intelligibility of Epistemic Activities. In De Regt, Leonelli and Eigner (eds.), 64–82.

2009b. Operationalism. In *Stanford Encyclopedia of Philosophy*. https://plato.stanford.edu/archives/fall2009/entries/operationalism.

2011a. The Philosophical Grammar of Scientific Practice. *International Studies in the Philosophy of Science* 25: 205–21.

2011b. Compositionism as a Dominant Way of Knowing in Modern Chemistry. *History of Science* 49: 247–68.

2012a. *Is Water H₂O? Evidence, Realism and Pluralism*. Dordrecht: Springer.

2012b. Practicing Eighteenth-Century Science Today. In Mario Biagioli and Jessica Riskin (eds.), *Nature Engaged: Science in Practice from the Renaissance to the Present*, 41–58. New York: Palgrave Macmillan.

2014. Epistemic Activities and Systems of Practice: Units of Analysis in Philosophy of Science after the Practice Turn. In Soler et al. (eds.), 67–79.

2016a. The Rising of Chemical Natural Kinds through Epistemic Iteration. In Kendig (ed.), 33–46.

2016b. Pragmatic Realism. *Revista de Humanidades de Valparaiso* 4: 107–22.

2017a. Epistemic Iteration and Natural Kinds: Realism and Pluralism in Taxonomy. In Kenneth S. Kendler and Josef Parnas (eds.), *Philosophical Issues in Psychiatry*, vol IV: *Classification of Psychiatric Illnesses*, 229–45. Oxford University Press.

2017b. Operational Coherence as the Source of Truth. *Proceedings of the Aristotelian Society* 117: 103–22.

2017c. Operationalism: Old Lessons and New Challenges. In Nicola Mößner and Alfred Nordmann (eds.), *Reasoning in Measurement*, 25–38. London and New York: Routledge.

2017d. What History Tells Us about the Distinct Nature of Chemistry. *Ambix* 64: 360–74.

2017e. Prospects for an Integrated History and Philosophy of Composition. In Hannes Leitgeb et al. (eds.), *Logic, Methodology and Philosophy of Science – Proceedings of the 15th International Congress*, 215–31. London: College Publications.

2018. Is Pluralism Compatible with Scientific Realism? In Saatsi (ed.), 176–86.

2020a. Pragmatism, Perspectivism and the Historicity of Science. In Massimi and McCoy (eds.), 10–27.

2020b. Relativism, Perspectivism and Pluralism. In Kusch (ed.), 398–406.

2021. The Coherence of Feyerabend's Pluralist Realism. In Karim Bschir and Jamie Shaw (eds.), *Interpreting Feyerabend: Critical Essays*, 40–56. Cambridge University Press.

Chang, Hasok and Fisher, Grant. 2011. What the Ravens Really Teach Us: The Inherent Contextuality of Evidence. In William Twining, Philip Dawid and Mimi Vasilaki (eds.), *Evidence, Inference and Enquiry*, 341–66. Oxford University Press and the British Academy.

Chang, Hasok and Leonelli, Sabina. 2005. Infrared Metaphysics: The Elusive Ontology of Radiation (Part 1); Infrared Metaphysics: Radiation and Theory-Choice (Part 2). *Studies in History and Philosophy of Science* 36: 477–508; 686–705.

Chang, Hasok and Yi, Sang Wook. 2005. The Absolute and Its Measurement: William Thomson on Temperature. *Annals of Science* 62: 281–308.

Chirimuuta, Mazviita. 2015. *Outside Color: Perceptual Science and the Puzzle of Color in Philosophy*. Cambridge, MA: MIT Press.

Clark, Peter and Hale, Bob (eds.). 1994. *Reading Putnam*. Oxford: Blackwell.

Craig, Edward. 1990. *Knowledge and the State of Nature*. Oxford University Press.

Crombie, Alistair. 1961. Quantification in Medieval Physics. In Harry Woolf (ed.), *Quantification: A History of the Meaning of Measurement in the Natural and Social Sciences*, 13–30. Indianapolis: Bobbs-Merrill.

Curiel, Erik. (forthcoming). Why Rigid Designation Cannot Stand on Scientific Ground.

Daly, C. B. 1968. Polanyi and Wittgenstein. In Thomas A. Langford and William H. Poteat (eds.), *Intellect and Hope: Essays in the Thought of Michael Polanyi*, 136–68. Durham, NC: Duke University Press.

Dancy, Jonathan. 1985. *Introduction to Contemporary Epistemology*. Oxford: Blackwell.

Darwall, Stephen. 2006. *The Second-Person Standpoint: Morality, Respect, and Accountability*. Cambridge, MA: Harvard University Press.

David, Marian. 2016. The Correspondence Theory of Truth. In *Stanford Encyclopedia of Philosophy*. https://plato.stanford.edu/archives/fall2016/entries/truth-correspondence.

Dawid, Richard. 2013. *String Theory and the Scientific Method*. Cambridge University Press.

De Regt, Henk W. 2017. *Understanding Scientific Understanding*. New York: Oxford University Press.

De Regt, Henk W., Leonelli, Sabina and Eigner, Kai (eds.). 2009. *Scientific Understanding: Philosophical Perspectives*. University of Pittsburgh Press.

Debus, Allen G. 1967. Fire Analysis and the Elements in the Sixteenth and the Seventeenth Centuries. *Annals of Science* 23(2): 127–47.

Deleuze, Gilles and Guattari, Félix. [1980] 1987. *A Thousand Plateaus: Capitalism and Schizophrenia*, trans. Brian Massumi. Minneapolis: University of Minnesota Press.

DeVries, Willem A. 2005. *Wilfrid Sellars*. Chesham: Acumen.

Dewey, John. [1907] 1977. The Intellectual Criterion for Truth. In *The Collected Works of John Dewey: The Middle Works*, vol. IV, 50–75. Carbondale: Southern Illinois University Press.

1917. The Need for a Recovery of Philosophy. In John Dewey (ed.), *Creative Intelligence: Essays in the Pragmatic Attitude*, 3–69. New York: Holt.

1925. *Experience and Nature*. Ithaca, NY: Cornell University Press.

1929. *The Quest for Certainty: A Study of the Relation of Knowledge and Action*. New York: Minton, Balch and Co.

1938. *Logic: The Theory of Inquiry*. New York: Holt, Reinhardt & Winston.

Diamond, Cora. 1991. *The Realistic Spirit: Wittgenstein, Philosophy, and the Mind*. Cambridge, MA: MIT Press.

Douglas, Heather. 2009. *Science, Policy and the Value-Free Ideal*. University of Pittsburgh Press.

Douven, Igor. 2021. Abduction. In *Stanford Encyclopedia of Philosophy*. https://plato.stanford.edu/archives/sum2021/entries/abduction.

Dummett, Michael. 1981. *Frege: Philosophy of Language*. 2nd edn. London: Duckworth.

Dupré, John. 1993. *The Disorder of Things: Metaphysical Foundations of the Disunity of Science*. Cambridge, MA: Harvard University Press.

Dutilh Novaes, Catarina. 2020. Carnapian Explication and Ameliorative Analysis: A Systematic Comparison. *Synthese* 197: 1011–34.

Eddington, Arthur S. 1928. *The Nature of the Physical World*. Cambridge University Press.

Einstein, Albert. 1961. *Relativity: The Special and the General Theory*, trans. Robert W. Lawson. New York: Crown.

Einstein, Albert and Infeld, Leopold. 1938. *The Evolution of Physics*. Cambridge University Press.

Elgin, Catherine Z. 2017. *True Enough*. Cambridge, MA: MIT Press.

Epstein, Brian. 2015. *The Ant Trap: Rebuilding the Foundations of the Social Sciences*. New York: Oxford University Press.

Faraday, Michael. 1822. On Some New Electro-Magnetic Motions, and on the Theory of Magnetism. *Quarterly Journal of Science, Literature and the Arts* 12: 74–96.

Fernflores, Francisco. 2012. The Equivalence of Mass and Energy. In *Stanford Encyclopedia of Philosophy*. http://plato.stanford.edu/archives/spr2012/entries/equivME.

Fesmire, Steven. 2015. *Dewey*. London and New York: Routledge.

Feyerabend, Paul. 1975. *Against Method*. London: New Left Books.

Fine, Arthur. 1984. The Natural Ontological Attitude. In Jarrett Leplin (ed.), *Scientific Realism*, 83–107. Berkeley and Los Angeles: University of California Press.

2007. Relativism, Pragmatism, and the Practice of Science. In Misak (ed.), 50–67.

Floyd, Juliet and Shieh, Sanford (eds.). 2001. *Future Pasts: The Analytic Tradition in Twentieth-Century Philosophy*. Oxford University Press.

Foley, Richard. 1998. Justification, Epistemic. In Edward Craig (ed.), *Routledge Encyclopedia of Philosophy*, 157–65. London: Routledge.

Føllesdal, Dagfinn. 1990. The *Lebenswelt* in Husserl. In Leila Haaparanta, Martin Kusch and Ilkka Niiniluoto (eds.), *Language, Knowledge, and Intentionality: Perspectives on the Philosophy of Jaakko Hintikka*. *Acta Philosophica Fennica* 46: 123–43.

2010. The *Lebenswelt* in Husserl. In David Hyder and Hans-Jörg Rheinberger (eds.), *Science and the Life-World: Essays on Husserl's 'Crisis of European Sciences'*, 27–45. Stanford University Press.

Forman, Paul. 1991. Independence, not Transcendence, for the Historian of Science. *Isis* 82: 71–86.

Frank, Philipp. 1949. *Modern Science and Its Philosophy*. Cambridge, MA: Harvard University Press.

Frega, Roberto (ed.) 2011. *Pragmatist Epistemologies*. Lanham, MD: Lexington.

Frege, Gottlob. [1892] 1948. Sense and Reference. *Philosophical Review* 57: 209–30.

Friedman, Michael. 2001. *Dynamics of Reason*. Stanford, CA: CSLI Publications.

Frigg, Roman and Nguyen, James. 2020. *Modelling Nature*. Cham: Springer Nature.

Gabriel, Markus. [2013] 2015. *Why the World Does Not Exist*. Cambridge: Polity Press.

Galison, Peter. 1997. *Image and Logic: A Material Culture of Microphysics*. University of Chicago Press.

2003. *Einstein's Clocks, Poincaré's Maps: Empires of Time*. New York: W. W. Norton.

Gardner, Martin. 2000. *Did Adam and Eve Have Navels?* New York: W. W. Norton.

Giere, Ronald N. 2004. How Models Are Used to Represent Reality. *Philosophy of Science* 71: 742–52.

2006. *Scientific Perspectivism*. University of Chicago Press.

2016. Feyerabend's Perspectivism. *Studies in History and Philosophy of Science* 57: 137–41.

Glanzberg, Michael. 2018a. Truth. In *Stanford Encyclopedia of Philosophy*. https://plato.stanford.edu/archives/fall2018/entries/truth.

(ed.). 2018b. *The Oxford Handbook of Truth*. Oxford University Press.

Goldman, Alvin and Beddor, Bob. 2016. Reliabilist Epistemology. In *Stanford Encyclopedia of Philosophy*. https://plato.stanford.edu/archives/win2016/entries/reliabilism.

Gómez, Juan Carlos. 2004. *Apes, Monkeys, Children, and the Growth of Mind*. Cambridge, MA: Harvard University Press.

Gooding, David. 1990. *Experiment and the Making of Meaning: Human Agency in Scientific Observation*. Dordrecht: Kluwer.

Goodman, Nelson. 1978. *Ways of Worldmaking*. Indianapolis: Hackett.

Gottwald, Siegfried. 2020. Many-Valued Logic. In *Stanford Encyclopedia of Philosophy*. https://plato.stanford.edu/archives/sum2020/entries/logic-manyvalued.

Gould, Stephen Jay and Lewontin, Richard C. 1979. The Spandrels of San Marco and the Panglossian Paradigm: A Critique of the Adaptationist Programme. *Proceedings of the Royal Society of London* B205 (1161): 581–98.

Grene, Marjorie. 1974. *The Knower and the Known*. Berkeley and Los Angeles: University of California Press.

1987. Historical Realism and Contextual Objectivity: A Developing Perspective in the Philosophy of Science. In Nancy J. Nersessian (ed.), *The Process of Science: Contemporary Philosophical Approaches to Understanding Scientific Practice*, 69–81. Dordrecht: Martinus Nijhoff.

Grimm, Stephen R. (ed.). 2018. *Making Sense of the World: New Essays on the Philosophy of Understanding*. New York: Oxford University Press.

Grimm, Stephen R., Baumberger, Christoph and Ammon, Sabine (eds.). 2017. *Explaining Understanding: New Perspectives from Epistemology and Philosophy of Science*. New York and Abingdon: Routledge.

Haack, Susan. 2014. Do Not Block the Way of Inquiry. *Transactions of the Charles S. Peirce Society* 50: 319–39.

Hacking, Ian. 1983. *Representing and Intervening*. Cambridge University Press.

1989. Extragalactic Reality: The Case of Gravitational Lensing. *Philosophy of Science* 56: 555–81.

Hardcastle, Gary and Slater, Matthew. 2014. A Novel Exercise for Teaching the Philosophy of Science. *Philosophy of Science* 81: 1184–96.

Harré, Rom and Llored, Jean-Pierre. 2019. *The Analysis of Practices*. Newcastle upon Tyne: Cambridge Scholars.

Haslanger, Sally. 2000. Gender and Race: (What) Are They? (What) Do We Want Them To Be? *Noûs* 34: 31–55.

Held, Richard. 1965. Plasticity in Sensory-Motor Systems. *Scientific American* 213/5: 84–94.

Henne, Céline. 2022. Framed and Framing Inquiry: Development and Defence of John Dewey's Theory of Knowledge. PhD Dissertation, University of Cambridge.

Hentschel, Klaus. 2002. *Mapping the Spectrum: Techniques of Visual Representation in Research and Teaching*. Oxford University Press.

Hesse, Mary. 1977. Truth and the Growth of Scientific Knowledge. In Frederick Suppe and Peter D. Asquith (eds.), *PSA 1976, vol. 2 (Symposia)*, 261–81. East Lansing, MI: Philosophy of Science Association.

Hjortland, Ole Thomassen. 2017. Anti-exceptionalism about Logic. *Philosophical Studies* 174: 631–58.

Hoefer, Carl and Martí, Genoveva. 2020. Realism, Reference & Perspective. *European Journal for Philosophy of Science* 10, article 38: 1–22.

Hofmann, August. 1865. On the Combining Power of Atoms. *Notices of the Proceedings at the Meetings of the Members of the Royal Institution of Great Britain* 4: 401–30.

Holton, Gerald. 1969. Einstein, Michelson, and the 'Crucial' Experiment. *Isis* 60: 133–97.

Hookway, Christopher. 2004. Truth, Reality, and Convergence. In Cheryl Misak (ed.), *The Cambridge Companion to Peirce*, 127–49. Cambridge University Press.

Hornsby, Jennifer. 2004. Agency and Actions. In J. Hyman and H. Steward (eds.), *Agency and Action*, 1–23. Cambridge University Press.

2007. Knowledge and Abilities in Action. In Christian Kanzian and Edmund Runggaldier (eds.), *Cultures: Conflict – Analysis – Dialogue*, 165–80. Frankfurt: Ontos Verlag.

Horwich, Paul. 1998a. *Meaning*. Oxford: Clarendon Press.

1998b. *Truth*. 2nd edn. Oxford University Press.

Hossenfelder, Sabine. 2018. *Lost in Math: How Beauty Leads Physics Astray*. New York: Basic Books.

Hoyningen-Huene, Paul. 2013. *Systematicity: The Nature of Science*. New York: Oxford University Press.

Hoyningen-Huene, Paul and Sankey, Howard (eds.). 2001. *Incommensurability and Related Matters*. Dordrecht: Kluwer.

Husserl, Edmund. [1954] 1970. *The Crisis of European Sciences and Transcendental Phenomenology*. Evanston, IL: Northwestern University Press.

Hyman, John. 1999. How Knowledge Works. *Philosophical Quarterly* 49: 433–51.

Ichikawa, Jonathan Jenkins and Steup, Matthias. 2018. The Analysis of Knowledge. In *Stanford Encyclopedia of Philosophy*. https://plato.stanford.edu/archives/sum2018/entries/knowledge-analysis.

James, William. [1907] 1975. *Pragmatism*. Cambridge, MA: Harvard University Press.

Kant, Immanuel. [1786] 2004. *Metaphysical Foundations of Natural Science*, ed. Michael Friedman. Cambridge University Press.

[1787] 1998. *The Critique of Pure Reason*. Cambridge University Press.

Kellert, Stephen H., Longino, Helen E. and Waters, C. Kenneth (eds.). 2006. *Scientific Pluralism*. Minneapolis: University of Minnesota Press.

Kendig, Catherine (ed.). 2016. *Natural Kinds and Classification in Scientific Practice*. Abingdon and New York: Routledge.

Kenny, Anthony J. P. 1989. *The Metaphysics of Mind*. Oxford University Press.

Khlentzos, Drew. 2021. Challenges to Metaphysical Realism. In *Stanford Encyclopedia of Philosophy*. https://plato.stanford.edu/archives/spr2021/entries/realism-sem-challenge.

Kim, Sung Ho. 2019. Max Weber. In *Stanford Encyclopedia of Philosophy*. https://plato.stanford.edu/archives/win2019/entries/weber.

Kirkham, R. L. 1992. *Theories of Truth: A Critical Introduction*. Cambridge, MA: MIT Press.

Kitcher, Philip. 1993. *The Advancement of Science: Science without Legend, Objectivity without Illusions*. New York and Oxford: Oxford University Press.

2001. *Science, Truth and Democracy*. New York: Oxford University Press.

2011a. *Science in a Democratic Society*. Amherst, NY: Prometheus Books.

2011b. Epistemology without History Is Blind. *Erkenntnis* 75: 505–24.

2012. *Preludes to Pragmatism: Toward a Reconstruction of Philosophy*. New York: Oxford University Press.

(forthcoming). *Homo Quaerens: Progress, Truth, and Values*.

Kitcher, Philip and Keller, Evelyn Fox. 2017. *The Seasons Alter: How to Save Our Planet in Six Acts*. New York: Liveright.

Klein, Ursula. 2003. *Experiments, Models, Paper Tools: Cultures of Organic Chemistry in the Nineteenth Century*. Stanford University Press.

Knorr Cetina, Karin. 2001. Objectual Practice. In Schatzki, Knorr Cetina and von Savigny (eds.), 186–97.

Kosso, Peter. 1998. *Appearance and Reality: An Introduction to the Philosophy of Physics*. New York and Oxford: Oxford University Press.

Kripke, Saul. 1980. *Naming and Necessity*. Cambridge, MA: Harvard University Press.

Kuhn, Thomas S. 1957. *The Copernican Revolution*. Cambridge, MA: Harvard University Press.

[1962] 1970. *The Structure of Scientific Revolutions*. 2nd edn. University of Chicago Press.

[1974] 1977. Second Thoughts on Paradigms. In Kuhn, *The Essential Tension*, 293–319. University of Chicago Press.

[1989] 2000. Possible Worlds in History of Science. In James Conant and John Haugeland (eds.), *The Road Since Structure*, 58–89. University of Chicago Press.

Kulp, Christopher B. 2009. Dewey, the Spectator Theory of Knowledge, and Internalism/Externalism. *The Modern Schoolman* 87/1: 67–77.

Kusch, Martin. 2002. *Knowledge by Agreement: The Programme of Communitarian Epistemology*. Oxford University Press.

(ed.). 2020. *The Routledge Handbook of Philosophy of Relativism*. London and New York: Routledge.

Ladyman, James and Ross, Don, with Spurrett, David, and Collier, John. 2007. *Every Thing Must Go: Metaphysics Naturalized*. Oxford University Press.

Lakatos, Imre and Musgrave, Alan (eds.). 1970. *Criticism and the Growth of Knowledge*. Cambridge University Press.

Lakoff, George and Johnson, Mark. 1980. *Metaphors We Live By*. University of Chicago Press.

LaPorte, Joseph. 2018. Rigid Designators. In *Stanford Encyclopedia of Philosophy*. https://plato.stanford.edu/archives/spr2018/entries/rigid-designators.

Laudan, Larry. 1977. *Progress and Its Problems*. Berkeley and Los Angeles: University of California Press.

1981. A Confutation of Convergent Realism. *Philosophy of Science* 48: 19–49.

1984. *Science and Values*. Berkeley and Los Angeles: University of California Press.

Lawler, Insa. 2018. Knowing Why – An Investigation of Explanatory Knowledge. PhD dissertation, University of Duisburg–Essen.

Legg, Catherine and Hookway, Christopher. 2021. Pragmatism. In *Stanford Encyclopedia of Philosophy*. https://plato.stanford.edu/archives/sum2021/entries/pragmatism.

Lehrer, Keith. 1990. *Theory of Knowledge*. Boulder, CO, and San Francisco, CA: Westview Press.

Lenoir, Timothy. 1992. Practical Reason and the Construction of Knowledge: The Life-world of Haber–Bosch. In Ernan McMullin (ed.), *The Social Dimensions of Science*, 158–97. University of Notre Dame Press.

Leonelli, Sabina. 2016. *Data-Centric Biology: A Philosophical Study*. University of Chicago Press.

Lewis, Clarence Irving. 1929. *Mind and the World-Order: Outline of a Theory of Knowledge*. New York: Dover.

1930. [Review of] The Quest for Certainty: A Study of the Relation of Knowledge and Action [by] John Dewey. *Journal of Philosophy* 27: 14–25.

Lõhkivi, Endla and Vihalemm, Rein (eds.). 2012. *Towards a Practical Realist Account of Science*, special issue of *Studia Philosophica Estonica* 5/2.

Longino, Helen E. 1990. *Science as Social Knowledge: Values and Objectivity in Scientific Inquiry*. Princeton University Press.

(forthcoming). What's Social about Social Epistemology? *Journal of Philosophy*.

Losee, John. 1993. *A Historical Introduction to the Philosophy of Science*. 3rd edn. Oxford and New York: Oxford University Press.

Lowry, T. M. 1936. *Historical Introduction to Chemistry*. London: Macmillan.

Lynch, Michael P. 2001. A Functionalist Theory of Truth. In Michael P. Lynch (ed.), *The Nature of Truth: Classical and Contemporary Perspectives*, 723–49. Cambridge, MA: MIT Press.

2009. *Truth as One and Many*. Oxford University Press.

Lyons, Timothy D. 2003. Explaining the Success of a Scientific Theory. *Philosophy of Science* 70: 891–901.

2016a. Scientific Realism. In Paul Humphreys (ed.), *The Oxford Handbook of Philosophy of Science*, 564–84. Oxford University Press.

2016b. Structural Realism versus Deployment Realism: A Comparative Evaluation. *Studies in History and Philosophy of Science, Part A* 59: 95–105.

Macarthur, David. 2012. Putnam and the Philosophical Appeal to Common Sense. In Baghramian (ed.), 127–41.

Mach, Ernst. [1889] 2013. *The Science of Mechanics: A Critical and Historical Exposition of Its Principles*, trans. Thomas J. McCormack. Cambridge University Press.

Mäki, Uskali. 2009. Realistic Realism about Unrealistic Models. In Don Ross and Harold Kincaid (eds.), *The Oxford Handbook of Philosophy of Economics*, 68–98. Oxford University Press.

Massimi, Michela. 2018a. Perspectivism. In Saatsi (ed.), 164–75.

2018b. Four Kinds of Perspectival Truth. *Philosophy and Phenomenological Research* 96: 342–59.

2021. Realism, Perspectivism, and Disagreement in Science. *Synthese* 198: 6115–41.

Massimi, Michela and McCoy, Casey. (eds.). 2020. *Understanding Perspectivism*. New York and London: Routledge.

Maudlin, Tim. 2015. Confessions of a Hardcore, Unsophisticated Metaphysical Realist. In Auxier et al. (eds.), 487–501.

Maxwell, Nicholas. 1984. *From Knowledge to Wisdom: A Revolution in the Aims and Methods of Science*. Oxford: Basil Blackwell.

Mayo, Deborah G. 1996. *Error and the Growth of Experimental Knowledge*. University of Chicago Press.

McIntyre, Lee. 2018. *Post-Truth*. Cambridge, MA: MIT Press.

2021. *How to Talk to a Science Denier*. Cambridge, MA: MIT Press.

McKenzie, Kerry. 2011. Arguing against Fundamentality. *Studies in History and Philosophy of Modern Physics* 42: 244–55.

2018. Being Realistic: The Challenge of Theory Change for a Metaphysics of Scientific Realism. *Spontaneous Generations* 9: 136–42.

McLaughlin, Amy. 2009. Peircean Polymorphism: Between Realism and Anti-realism. *Transactions of the Charles S. Peirce Society* 45: 402–21.

2011. In Pursuit of Resistance: Pragmatic Recommendations for Doing Science within One's Means. *European Journal for Philosophy of Science* 1: 353–71.

McLeod, Alexus. 2016. *Theories of Truth in Chinese Philosophy: A Comparative Approach*. London and New York: Rowman & Littlefield.

Merleau-Ponty, Maurice. [1945] 1962. *Phenomenology of Perception*. London and Henley: Routledge & Kegan Paul.

Michaelson, Eliot and Reimer, Marga. 2019. Reference. In *Stanford Encyclopedia of Philosophy*. https://plato.stanford.edu/archives/spr2019/entries/reference.

Miller, Alexander. 2016. Realism. In *Stanford Encyclopedia of Philosophy*. https://plato.stanford.edu/archives/win2016/entries/realism.

Miller, Arthur I. 1981. *Albert Einstein's Special Theory of Relativity: Emergence (1905) and Early Interpretation (1905–1911)*. Reading, MA: Addison-Wesley.

Misak, Cheryl (ed.). 2007a. *New Pragmatists*. Oxford: Clarendon Press.

2007b. Pragmatism and Deflationism. In Misak (ed.), 68–90.

2013. *The American Pragmatists*. Oxford University Press.

2020. *Frank Ramsey: A Sheer Excess of Powers*. Oxford University Press.

Mitchell, Sandra D. 2003. *Biological Complexity and Integrative Pluralism*. Cambridge University Press.

2020. Perspectives, Representation and Integration. In Massimi and McCoy, (eds.), 178–93.

Mol, Annemarie. 2002. *The Body Multiple: Ontology in Medical Practice*. Durham, NC: Duke University Press.

Moore, G. E. 1939. Proof of an External World. *Proceedings of the British Academy* 25: 273–300.

Morgan, Mary S. and Morrison, Margaret (eds.). 1999. *Models as Mediators: Perspectives on Natural and Social Science*. Cambridge University Press.

Murdoch, Dugald. 1987. *Niels Bohr's Philosophy of Physics*. Cambridge University Press.

Musgrave, Alan. 1976. Why Did Oxygen Supplant Phlogiston? Research Programmes in the Chemical Revolution. In Colin Howson (ed.), *Method and Appraisal in the Physical Sciences*, 181–209. Cambridge University Press.

Nersessian, Nancy J. 2008. *Creating Scientific Concepts*. Cambridge, MA: MIT Press.

Neurath, Otto. [1931] 1983. Sociology in the Framework of Physicalism. In Robert S. Cohen and Marie Neurath (eds.), Neurath, *Philosophical Papers 1913–1946*, 58–90. Dordrecht: Reidel.

[1932/3] 1983. Protocol Statements. In Robert S. Cohen and Marie Neurath (eds.), Neurath, *Philosophical Papers 1913–1946*, 91–9. Dordrecht: Reidel.

Neurath, Otto et al. [1929] 1973. Wissenschaftliche Weltauffassung: Der Wiener Kreis (The Scientific Conception of the World: The Vienna Circle). In Marie Neurath and Robert S. Cohen (eds.), Neurath, *Empiricism and Sociology*, 299–318. Dordrecht: Reidel.

Niiniluoto, Ilkka. 1999. *Critical Scientific Realism*. New York: Oxford University Press.

2014. Scientific Realism: Independence, Causation, and Abduction. In Westphal (ed.), 159–72.

Noë, Alva. 2004. *Action in Perception*. Cambridge, MA: MIT Press.

2005. Against Intellectualism. *Analysis* 65: 278–90.

Norton, John D. 2021. *The Material Theory of Induction.* University of Calgary Press.

Nye, Mary Jo. 1972. *Molecular Reality: A Perspective on the Scientific Work of Jean Perrin.* New York: American Elsevier.

Ogilvie, John F. 1990. The Nature of the Chemical Bond 1990: There Are No Such Things as Orbitals! *Journal of Chemical Education* 67: 280–9.

Oppenheim, Paul and Putnam, Hilary. 1958. Unity of Science as a Working Hypothesis. In Herbert Feigl, Michael Scriven and Grover Maxwell (eds.), *Concepts, Theories, and the Mind-Body Problem,* 3–36. Minneapolis: University of Minnesota Press.

Page, Sam. 2006. Mind-Independence Disambiguated: Separating the Meat from the Straw in the Realism/Anti-realism Debate. *Ratio* 19/3: 321–35.

Pedersen, Nikolaj Jang Lee Linding and Lynch, Michael P. 2018. Truth Pluralism. In Glanzberg (ed.), 543–78.

Pedersen, Nikolaj Jang Lee Linding and Wright, Cory. 2018. Pluralist Theories of Truth. In *Stanford Encyclopedia of Philosophy.* https://plato.stanford.edu/archives/win2018/entries/truth-pluralist.

Peirce, Charles Sanders. 1877. The Fixation of Belief. *Popular Science Monthly* 12/1: 1–15.

1878. How to Make Our Ideas Clear. *Popular Science Monthly* 12/3: 286–302.

1934. *Collected Papers of Charles Sanders Peirce,* vol. v. Cambridge, MA: Harvard University Press.

1986. *Writings of Charles S. Peirce: A Chronological Edition,* vol. III. Bloomington: Indiana University Press.

Peters, Dean. 2012. How to Be a Scientific Realist (If At All): A Study of Partial Realism. PhD dissertation, London School of Economics.

Piaget, Jean. [1937] 1954. *The Construction of Reality in the Child.* New York: Basic Books.

Pickering, Andrew. 1995. *The Mangle of Practice.* University of Chicago Press.

2015. Science, Contingency, and Ontology. In Léna Soler et al. (eds.), *Science as It Could Have Been: Discussing the Contingency/Inevitability Problem,* 117–28. University of Pittsburgh Press.

Pihlström, Sami. 2003. *Naturalizing the Transcendental: A Pragmatic View.* Amherst, NY: Prometheus.

2009. *Pragmatist Metaphysics: An Essay on the Ethical Grounds of Ontology.* London: Continuum.

2011. The Problem of Realism, from a Pragmatic Point of View. In Frega (ed.), 103–26.

2012. Toward Pragmatically Naturalized Transcendental Philosophy of Scientific Inquiry and Pragmatic Scientific Realism. *Studia Philosophica Estonica* 5/2: 79–94.

2014. Pragmatic Realism. In Westphal (ed.), 251–82.

Pirsig, Robert M. 1974. *Zen and the Art of Motorcycle Maintenance.* New York: William Morrow.

Polanyi, Michael. 1958. *Personal Knowledge: Towards a Post-critical Philosophy.* University of Chicago Press.

Price, Huw. 1988. *Facts and the Function of Truth.* Oxford: Basil Blackwell.

1998. Three Norms of Assertibility, or How the Moa became Extinct. *Philosophical Perspectives* 12: 241–54.

2003. Truth as Convenient Friction. *Journal of Philosophy* 100/4: 167–90.

Priest, Graham. 2008. *An Introduction to Non-classical Logic: From If to Is.* 2nd edn. Cambridge University Press.

Pritchard, Duncan. 2016. Seeing It for Oneself: Perceptual Knowledge, Understanding, and Intellectual Autonomy. *Episteme* 13/1: 29–42.

Privat-Deschanel, Agustin. 1876. *Elementary Treatise on Natural Philosophy,* trans. and ed. J. D. Everett. London: Blackie & Son.

Psillos, Stathis. 1999. *Scientific Realism: How Science Tracks Truth.* London and New York: Routledge.

Putnam, Hilary. 1975a. *Philosophical Papers, vol. 1: Mathematics, Matter and Method.* Cambridge University Press.

1975b. What Is Mathematical Truth? In Putnam 1975a, 60–78.

1977. Realism and Reason. *Proceedings and Addresses of the American Philosophical Association* 50/6: 483–98.

1980a. How to Be an Internal Realist and a Transcendental Idealist (at the Same Time). In R. Haller and W. Grassl (eds.), *Language, Logic, and Philosophy,* 100–8. Vienna: Hölder-Pichler-Tempsky.

1980b. Models and Reality. *Journal of Symbolic Logic* 45: 464–82.

1981. *Reason, Truth and History.* Cambridge University Press.

1982. Why There Isn't a Ready-Made World. *Synthese* 51: 141–67.

1983. *Realism and Reason.* New York: Cambridge University Press.

1987. *The Many Faces of Realism (The Paul Carus Lectures).* Chicago and La Salle, IL: Open Court.

1990a. *Realism with a Human Face,* ed. James Conant. Cambridge, MA: Harvard University Press.

1990b. Realism with a Human Face. In Putnam 1990a, 3–29.

1990c. A Defense of Internal Realism. In Putnam 1990a, 30–42.

1995. *Pragmatism: An Open Question.* Oxford: Blackwell.

1999. *The Threefold Cord: Mind, Body and World.* New York: Columbia University Press.

2015a. Intellectual Autobiography. In Auxier et al. (eds.), 1–110.

2015b. Reply to Tim Maudlin. In Auxier et al. (eds.), 502–9.

Quine, W. V. O. 1969. *Ontological Relativity and Other Essays.* New York: Columbia University Press.

Quinn, Terry. 2011. *From Artefacts to Atoms.* Oxford University Press.

Radder, Hans. 2006. *The World Observed / The World Conceived.* University of Pittsburgh Press.

Raskin, Marcus G. and Bernstein, Herbert J. 1987. *New Ways of Knowing: The Sciences, Society and Reconstructive Knowledge.* Totowa, NJ: Rowman & Littlefield.

Rasmussen, Nicolas. 1993. Facts, Artifacts, and Mesosomes: Practicing Epistemology with the Electron Microscope. *Studies in History and Philosophy of Science* 24: 227–65.

Ray, Greg. 2018. Tarski on the Concept of Truth. In Glanzberg (ed.), 695–717.

Reichenbach, Hans. [1920] 1965. *The Theory of Relativity and A Priori Knowledge.* Berkeley and Los Angeles: University of California Press.

Rescher, Nicholas. 1980. Conceptual Schemes. *Midwest Studies in Philosophy* 5: 323–46.

Resnick, David. 1994. Hacking's Experimental Realism. *Canadian Journal of Philosophy* 24: 395–412.

Rorty, Richard. 1982. *Consequences of Pragmatism: Essays 1972–1980.* Minneapolis: University of Minnesota Press.

Rosenthal, Sandra B. 2007. *C. I. Lewis in Focus: The Pulse of Pragmatism.* Bloomington: Indiana University Press.

Rouse, Joseph. 1987. *Knowledge and Power: Toward a Political Philosophy of Science.* Ithaca, NY: Cornell University Press.

2001. Two Concepts of Practices. In Schatzki, Knorr Cetina and von Savigny (eds.), 198–208.

2015. *Articulating the World: Conceptual Understanding and the Scientific Image.* University of Chicago Press.

Rowbottom, Darrell P. 2019. *The Instrument of Science: Scientific Anti-realism Revitalized.* London and New York: Routledge.

Ruphy, Stéphanie. 2016. *Scientific Pluralism Reconsidered: A New Approach to the (Dis)Unity of Science.* University of Pittsburgh Press.

Russell, Bertrand. 1910. William James's Conception of Truth. In Russell, *Philosophical Essays,* 127–49. London: Longmans, Green.

1912. *Problems of Philosophy.* London: Williams and Norgate.

Ruthenberg, Klaus and Chang, Hasok. 2020. Glass and Life: The Biochemical Origins of pH. *Beitrag für die Mitteilungen der Fachgruppe Geschichte der Chemie* 26: 63–87.

Ryle, Gilbert. 1946. Knowing How and Knowing That: The Presidential Address. *Proceedings of the Aristotelian Society,* new series 46: 1–16.

Saatsi, Juha (ed.). 2018. *The Routledge Handbook of Scientific Realism.* London and New York: Routledge.

Sagan, Carl. 1980. *Cosmos.* New York: Random House.

Scerri, Eric. 2007. *The Periodic Table: Its Story and Its Significance.* Oxford University Press.

Scharp, Kevin. 2013. *Replacing Truth.* Oxford University Press.

2021. Conceptual Engineering for Truth: Alethic Properties and New Alethic Concepts. *Synthese* 198: 647–88.

Schatzki, Theodore R. 2001a. Introduction: Practice Theory. In Schatzki, Knorr Cetina and von Savigny (eds.), 10–23.

2001b. Practice Mind-ed Orders. In Schatzki, Knorr Cetina and von Savigny (eds.), 50–63.

Schatzki, Theodore R., Knorr Cetina, Karin and von Savigny, Eike (eds.). 2001. *The Practice Turn in Contemporary Theory*. London and New York: Routledge.

Scheffler, Israel. 1999. A Plea for Pluralism. *Transactions of the Charles S. Peirce Society* 35: 425–36.

Schiller, Ferdinand Canning Scott. 1939. *Our Human Truths*. New York: Columbia University Press.

Schilpp, Paul Arthur (ed.). 1968. *The Philosophy of C. I. Lewis*. La Salle, IL: Open Court.

Schweikard, David P. and Schmid, Hans Bernhard. 2021. Collective Intentionality. In *Stanford Encyclopedia of Philosophy*. https://plato.stanford.edu/archives/fall2021/entries/collective-intentionality.

Šešelja, Dunja, Kosolosky, Laszlo and Straßer, Christian. 2012. The Rationality of Scientific Reasoning in the Context of Pursuit: Drawing Appropriate Distinctions. *Philosophica* 86: 51–82.

Šešelja, Dunja and Straßer, Christian. 2014. Epistemic Justification in the Context of Pursuit: A Coherentist Approach. *Synthese* 191: 3111–41.

Shan, Yafeng. 2019. A New Functional Approach to Scientific Progress. *Philosophy of Science* 104: 639–59.

Shang, Rick. 2021. Positron Emission Tomography from 1930 to 1990. PhD dissertation, Washington University.

Shapere, Dudley. 1993. Discussion: Astronomy and Anti-Realism. *Philosophy of Science* 60:.134–50.

Shipley, Joseph T. 1984. *The Origins of English Words: A Discursive Dictionary of Indo-European Roots*. Baltimore, MD: Johns Hopkins University Press.

Sismondo, Sergio. 2017. Post-Truth? *Social Studies of Science* 46/1: 3–6.

Skulberg, Emilie. 2021. The Event Horizon as a Vanishing Point. PhD dissertation, University of Cambridge.

Smart, J. J. C. 1963. *Philosophy and Scientific Realism*. London: RKP.

Snowdon, Paul. 2004. The Presidential Address. Knowing How and Knowing That: A Distinction Reconsidered. *Proceedings of the Aristotelian Society*, new series 104: 1–32.

Soler, Léna. 2009. *Introduction à l'épistémologie*. 2nd edn. Paris: Ellipses.

2012. Robustness of Results and Robustness of Derivations: The Internal Architecture of a Solid Experimental Proof. In Léna Soler, Emiliano Trizio, Thomas Nickles and William C. Wimsatt (eds.), *The Robustness of Science*, 227–66. Dordrecht: Springer.

Soler, Léna and Catinaud, Régis. 2014. Toward a Framework for the Analysis of Scientific Practices. In Soler et al. (eds.), 80–92.

Soler, Léna, Zwart, Sjoerd, Lynch, Michael and Israel-Jost, Vincent (eds.). 2014. *Science after the Practice Turn in the Philosophy, History and Social Studies of Science*. New York and London: Routledge.

Solomon, Miriam. 2001. *Social Empiricism*. Cambridge, MA: MIT Press.

Sosa, Ernest. 1993. Putnam's Pragmatic Realism. *Journal of Philosophy* 90: 605–26.

2017. *Epistemology*. Princeton University Press.

Spiegelberg, Herbert. 1956. Husserl and Peirce's Phenomenologies: Coincidence or Interaction. *Philosophy and Phenomenological Research* 17: 164–85.

Staley, Richard. 2008. *Einstein's Generation: The Origins of the Relativity Revolution.* University of Chicago Press.

Stanford, P. Kyle. 2018. Unconceived Alternatives and the Strategy of Historical Ostension. In Saatsi (ed.), 212–24.

Stanford, P. Kyle and Kitcher, Philip. 2000. Refining the Causal Theory of Reference for Natural Kind Terms. *Philosophical Studies* 97: 99–129.

Stang, Nicholas F. 2018. Kant's Transcendental Idealism. In *Stanford Encyclopedia of Philosophy.* https://plato.stanford.edu/archives/win2018/entries/kant-transcendental-idealism.

Stanley, Jason and Williamson, Timothy. 2001. Knowing How. *Journal of Philosophy* 98: 411–44.

Steinhoff, Gordon. 1986. Internal Realism, Truth and Understanding. In Arthur Fine and Peter Machamer (eds.), *PSA 1986*, vol. 1, 352–63. East Lansing, MI: Philosophy of Science Association.

Steinle, Friedrich. 2016. *Exploratory Experiments: Ampère, Faraday, and the Origins of Electrodynamics.* University of Pittsburgh Press.

Stern, David G. 2003. The Practical Turn. In Stephen P. Turner and Paul A. Roth (eds.), *The Blackwell Guide to the Philosophy of the Social Sciences*, 185–206. Oxford: Blackwell.

Stoljar, Daniel and Damnjanovic, Nic. 2014. The Deflationary Theory of Truth. In *Stanford Encyclopedia of Philosophy.* https://plato.stanford.edu/archives/fall2014/entries/truth-deflationary.

Stuart, Michael T. 2018. How Thought Experiments Increase Understanding. In Stuart, Fehige and Brown (eds.), 526–44.

Stuart, Michael T., Fehige, Yiftach and Brown, James Robert (eds.). 2018. *The Routledge Companion to Thought Experiments.* London and New York: Routledge.

Stump, David J. 2015. *Conceptual Change and the Philosophy of Science: Alternative Interpretations of the A Priori.* New York: Routledge.

2020. Ontological Relativity. In Kusch (ed.), 341–8.

Suárez, Mauricio. 2004. An Inferential Conception of Scientific Representation. *Philosophy of Science* 71: 767–79.

(ed.). 2009. *Fictions in Science.* New York: Routledge.

2016. Representation in Science. In Paul Humphreys (ed.), *The Oxford Handbook of Philosophy of Science*, 440–59. Oxford University Press.

Swoyer, Chris. 1991. Structural Representation and Surrogative Reasoning. *Synthese* 87: 449–508.

Teller, Paul. 2001. Twilight of the Perfect Model Model. *Erkenntnis* 55: 393–415.

2017. Modeling Truth. *Philosophia* 45: 143–61.

2018. Referential and Prospectival Realism. *Spontaneous Generations* 9/1: 151–64.

2020. What Is Perspectivism, and Does It Count as Realism? In Massimi and McCoy (eds.), 49–64.

2021. Making Worlds with Symbols. *Synthese* 198 (Suppl 21): S5015–36.

Thagard, Paul. 2000. *Coherence in Thought and Action*. Cambridge, MA: MIT Press.

Thompson, Michael. 2008. *Life and Action: Elementary Structures of Practice and Practical Thought*. Cambridge, MA: Harvard University Press.

Tomczyk, Hannah. 2022. Did Einstein Predict Bose–Einstein Condensation? *Studies in History and Philosophy of Science* 93: 30–8.

Toon, Adam. 2012. *Models as Make-Believe: Imagination, Fiction and Scientific Representation*. Basingstoke: Palgrave Macmillan.

Torretti, Roberto. 1990. *Creative Understanding*. University of Chicago Press.

2000. 'Scientific Realism' and Scientific Practice. In Evandro Agazzi and Massimo Pauri (eds.), *The Reality of the Unobservable*, 113–22. Dordrecht: Kluwer.

Tuomela, Raimo. 1985. *Science, Action, and Reality*. Dordrecht: Reidel.

Van Fraassen, Bas. 1980. *The Scientific Image*. Oxford: Clarendon Press.

2002. *The Empirical Stance*. New Haven, CT: Yale University Press.

2004. Précis of *The Empirical Stance*. *Philosophical Studies* 121: 127–32.

2008. *Scientific Representation: Paradoxes of Perspective*. Oxford University Press.

Vickers, Peter. 2017. Understanding the Selective Realist Defence against the PMI. *Synthese* 194: 3221–32.

Vihalemm, Rein. 2012. Practical Realism: Against Standard Scientific Realism and Anti-realism. *Studia Philosophica Estonica* 5/2: 7–22.

Watkins, Calvert (ed.). 1985. *The American Heritage Dictionary of Indo-European Roots*. Boston: Houghton Mifflin.

Weinberg, Steven. 1992. *Dreams of a Final Theory*. New York: Random House.

Westerblad, Oscar. (forthcoming). Making Sense of Understanding: A Pragmatist Account of Scientific Understanding. PhD dissertation, University of Cambridge.

Westphal, Kenneth (ed.). 2014. *Realism, Science and Pragmatism*. Abingdon and New York: Routledge.

Wheaton, Bruce R. 1983. *The Tiger and the Shark: Empirical Roots of Wave–Particle Dualism*. Cambridge University Press.

Williamson, Timothy. 2000. *Knowledge and Its Limits*. Oxford University Press.

Wilson, Mark. 2006. *Wandering Significance: An Essay on Conceptual Behaviour*. Oxford University Press.

Wimsatt, William C. 2007. *Re-engineering Philosophy for Limited Beings: Piecewise Approximations to Reality*. Cambridge, MA: Harvard University Press.

Winther, Rasmus. 2020. *When Maps Become the World*. University of Chicago Press.

Wittgenstein, Ludwig. 1922. *Tractatus Logico-Philosophicus*. London: Routledge & Kegan Paul.

1953. *Philosophical Investigations*. Oxford: Blackwell.

1969. *On Certainty (Über Gewissheit)*. New York: Harper.

Woodward, James. (forthcoming). Sketch of Some Themes for a Pragmatist Philosophy of Science. In Holly Andersen and Sandra Mitchell (eds.), *The Pragmatist Challenge*. Oxford University Press.

Worrall, John. 1989. Structural Realism: The Best of Both Worlds? *Dialectica* 43: 99–124.

Wray, K. Brad. 2018. *Resisting Scientific Realism*. Cambridge University Press.

Wright, Crispin. 1992. *Truth and Objectivity*. Cambridge, MA: Harvard University Press.

    1999. Truth: A Traditional Debate Reviewed. *Canadian Journal of Philosophy* 24: 31–74.

Young, James O. 2015. The Coherence Theory of Truth. In *Stanford Encyclopedia of Philosophy*. https://plato.stanford.edu/archives/fall2015/entries/truth-coherence.

Zambito, Pascal. 2019. 'Logic is a geometry of thinking': Space and Spatial Frameworks in Wittgenstein's Writings. PhD dissertation, University of Cambridge.

Index

a priori, 133, 137–9, 169, 220, 244
  operational, 244
abduction, 250
abundance, 215, 236
acid, 140, 145
action, standard story of, 28
active knowledge, 17–23, 63
active realism, 209
activist realism, 8, 208–16, 251
  served by pluralism, 237–9
activity, 15, 33, 41, 218
  analysis of, 38
  constitution of, 39
  representational, 113
activity-based analysis, 35
activity–principle pair, 135
aether, 54, 57, 246
aim-oriented adjustment, 25, 49, 53
aim-oriented coordination, 24
aims
  adjustment of, 48
  inherent, 36–7, 39, 57
  iterative development of, 246
  renunciation of, 246
Ampère, André-Marie, 53
ampliative inference, 215, 250
ampliative inquiry, 215, 250
Ankeny, Rachel, 15
anti-realist practices, 210
Arabatzis, Theodore, 99
Arrhenius, Svante, 141
atom, concept of, 242
atomism, 92, 110, 211, 246, 248
Austin, J. L., 87, 120, 171
axiom, 177

ball-and-stick molecular model, 112, 115
belief, 4, 19, 23, 28, 41, 48, 73, 106, 169, 184,
  189, 198–9, 202, 221, 224–5
Bergson, Henri, 62, 136
Berkeley, George, 61, 103

Big Bang, 89, 104
black hole, 124
Black, Joseph, 242
bog-walking, 240
Bohr, Niels, 193, 196
Bokulich, Alisa, 237–8
Boltzmann, Ludwig, 211
Boon, Mieke, 125
bootstrapping (in particle physics), 159
Bose–Einstein condensate, 188
box project, 80
Boyd, Richard, 145, 171
Boyle, Robert, 158
Brading, Katherine, 108, 113
Brandom, Robert, 174
Brexit, 9
Bridgman, Percy Williams, 3, 16, 35, 215
Brønsted, J. N., 141, 146
Brown, Matthew, 71
Buber, Martin, 31
Buddhism, 212
Button, Tim, 70, 84, 91, 227

caloric, 107, 184, 194
Cao, Tian Yu, 171
capabilities (or, capacities), 29, 34
Capps, John, 202
carbon dioxide, 189, 242
carbon monoxide, 189
carbon, tetravalency of, 112, 188
Carnap, Rudolf, 72, 122, 149, 220, 229
Carnot, Sadi, 248
Cartwright, Nancy, 64, 113, 130, 149, 237–8
cash-value, 198, 202
Cassirer, Ernst, 74
Catinaud, Régis, 38
cave, Plato's, 80, 82
central dogma, 111, 130
certainty, 26
Chakravartty, Anjan, 137, 149, 231
Chalmers, Alan, 64, 92

Chew, Geoffrey, 159
Chirimuuta, Mazviita, 61
circularity, 42, 104–5, 166
classical mechanics
   as a closed theory, 238
   formulations of, 230
classical vs. quantum mechanics, 87, 172, 188, 238
   semiclassical mechanics, 238
closed theory, 237
coherentism (in epistemology), 42
common sense, 216, 226
complementary science, 251
compositionism, 157
concept, 72, 77, 125, 129, 230, 243
concept-development, 136, 139–41
   by epistemic iteration, 241–4
conceptual engineering, 122, 129, 175
conceptual framework/scheme, 72–3, 118, 132, 149, 155, 223, 230
concept-validation, 142
confirmation, 181, 185
conservatism, as a feature of epistemic iteration, 240
constellations, 125, 147
constraint (by nature), 169
constructive empiricism. *See* empiricism, constructive
constructivism, 42, 124–5, 211, 218, 230–1
   blending with realism, 211
continuity of inquiry, 217
continuity of rules, 63
contradiction, 193, 195
convenience vs. truth, 190
convention, 107, 177
cookie-cutter metaphor, 81
coordination of activities, 38–40
coordination, aim-oriented, 40–8
coordination, problem of, 76, 104
copy theory of knowledge, 198
correspondence, 5, 70, 77, 79, 180, 200, 219, 222–4
   in practice, 111–14
   Putnam's notion of, 226
correspondence realism, 68–100
correspondentitis, 79
critical realism, 150
Crull, Elise, 108
Curiel, Erik, 99

Damnjanovic, Nic, 164
Darwall, Stephen, 31
Dawid, Richard, 211
definition, 16, 137, 143, 177
deflationism, 164–5, 176

Dewey, John, 20, 25, 46, 62–7, 74, 169–70, 198, 208, 212–13, 241, 244, 254
Diamond, Cora, 207
dinosaurs, 123
direct realism (or, natural realism), 228
disease, 154
disquotation, 81, 164, 219
DNA, 110–11, 169
Douven, Igor, 250
duck–rabbit, 196
Duhem, Pierre, 246
Dummett, Michael, 86
Dupré, John, 126, 156

earthquake, 168
Eddington, Arthur, 154
effective false beliefs, objection from, 189
effective theory, 171
Einstein, Albert, 54, 57, 88, 117, 137, 177, 195, 246
electromagnetism, 52
electron, 97, 99, 186, 233
Elgin, Catherine, 42, 171, 245
Emerson, Ralph Waldo, 61
empirical domain, 167
empiricism, 205, 252
   and activist realism, 248
   constructive, 69, 125, 205, 247
   relentless, 60, 252
   traditional, 61
Engels, Friedrich, 219
Enlightenment, 212
entity, 139
entity realism, 122, 249
epistemic activity, 16, 35–6, 86
epistemic agents, 19, 27–33
epistemic iteration. *See* iteration, epistemic
Epstein, Brian, 30
etymology of 'know' and 'can', 18
evening star and morning star, 96
existence, 146
experience, 61, 142, 225, 252
experimental realism, 120, 150, 247
explication, 122
exploratory experimentation, 53
extensionalism, 92–3
external realism, 71, 84
externalism, 219, 224

faith in science, 100–11
fallibilism, 76, 150
falsity, 192
Faraday, Michael, 53
Feyerabend, Paul, 102, 232, 236, 238
fictionalism, 186

Fine, Arthur, 173, 206
fire-analysis, 158
fixed air, 242
flat-earth theory, 190, 240
Fleck, Ludwig, 49
Føllesdal, Dagfinn, 32
force, 142
Forman, Paul, 210
foundationalism, 240
Fourier, Joseph, 248
framing, etymology of, 132
Frege, Gottlob, 96
Friedman, Michael, 137, 139, 220, 244
Frigg, Roman, 114
frigorific radiation, 151
function, external, 36

Gabriel, Markus, 128
Galilean strategy, 222
Galilei, Galileo, 64
Galileo's law of free fall, 250
Galison, Peter, 41
geocentrism, 234
Giere, Ronald, 81, 113, 208, 229–31
glasses in the mirror, 241
Global Positioning System (GPS), 4, 41, 239
God, 77, 86–7, 89, 128, 189, 201, 217, 252
God's Eye point of view, 7, 71, 74, 118, 217, 224, 232
gold, 93, 95–6
Gómez, Juan Carlos, 22, 95
Gooding, David, 53
Goodman, Nelson, 128, 138, 229, 233–4
Gosse, Philip Henry, 85
gravity, 110–11, 179, 187, 238
Grene, Marjorie, 5, 18, 133

Haack, Susan, 214
Hacking, Ian, 4, 13, 61, 94, 113, 120, 122, 124, 130, 150, 153, 227, 247, 249
Hales, Stephen, 242
Hardcastle, Gary, 80
hardness, 191
harmony, 43, 47
Heisenberg, Werner, 237
heliocentrism, 102
Henne, Céline, 26
hermeneutic dimension, 24, 36
Herschel, William, 151
Hesse, Mary, 105
high-energy experiments, 159, 187
Hoefer, Carl, 97
Hofmann, August, 112, 115
Hookway, Christopher, 58, 201
Hornsby, Jennifer, 28–9, 33

Horwich, Paul, 164, 202
Hossenfelder, Sabine, 211
Hoyningen-Huene, Paul, 18
Hughes, R. I. G., 113
humanism, 62, 228, 252
humility, 253
Husserl, Edmund, 28, 32, 61
hylomorphism, 157

ideal type, 36
idealization, 229
identity of indiscernibles, principle of, 135
inaccessibility argument, 123
incoherence, 41, 46, 181
incommensurability, 98, 195–6, 230
    semantic, 195
individualism, 31
individuation, 31
inductivism, 235
inevitability, 201
Infeld, Leopold, 117
inference from success to truth, 234
inference, everyday meaning of, 250
infrared radiation, 151
inquiry, 24–6, 48–57, 63, 213, 240, 243
    ampliative, 250
    end of, 198
    iterative, 244
    restricted vs. unrestricted, 50
    unrestricted, 56, 103
instrumentalism, 69, 186, 192, 204
intelligibility, 134
intercellular matrix, 160
internal realism, 91, 207, 219–29, 232
    Putnam's renunciation of, 228
internalism, 207, 219, 223
interpretation, 233
invention, 125
ism, meaning of, 208
iteration
    in building of coherence, 143
    in development of aims, 246
    in emergence of sociality, 30–3
    epistemic, 208, 239–47
    in evolution of meaning, 99
    mathematical, 241
    as a mode of progress, 208

James, William, 20, 62–7, 181–2, 192, 225, 228
    on truth, 198–203
Johnson, Mark, 81
Johnson, Samuel, 103
justification, 166

Kant, Immanuel, 62, 74, 79, 90, 133, 137, 150–1, 179, 208, 219–21, 225, 232
Keller, Evelyn Fox, 2
Kellert, Stephen, 7
Kelvin, Lord (William Thomson), 241
Kendig, Catherine, 78
Kenny, Anthony, 18
kilogram, 99
Kim, Sung Ho, 36
kinetic theory of gases, 194
Kitcher, Philip, 2, 17, 24, 59, 74, 98, 106, 150, 182, 207, 212–13, 221–3
Klein, Ursula, 35
Knorr Cetina, Karin, 34
knowledge, value of, 209
knowledge-as-ability, 18, 34
knowledge-as-information, 17
knowledge-by-acquaintance, 21, 95
Kosolosky, Laszlo, 246
Kosso, Peter, 7, 207
Kripke, Saul, 94, 96, 98, 100
Kuhn, Thomas, 14, 26, 72, 98, 105, 118, 195–6, 214, 220, 229
    on normal vs. extraordinary science, 50
Kusch, Martin, 30, 47

Ladyman, James, 157
Lakatos, Imre, 169
Lakoff, George, 81
Landry, Elaine, 113
Laplace, Pierre-Simon, 184
Laudan, Larry, 14, 98, 105
Lavoisier, Antoine-Laurent, 16, 141, 145, 194, 243
learning, 7, 22, 25, 29, 44, 49, 60, 62, 129, 204–5, 209, 215, 235, 247–8, 252
    from the process of learning, 67
    of methods, 63
Lego, invention of, 160
Legoism, 104, 160–2
    sources of, 160
Lehrer, Keith, 17, 23
Leibniz, G. W. von, 135
Lenoir, Timothy, 216
Leonelli, Sabina, 15, 130, 151
Lewis, Clarence Irving, 62, 66, 127, 137–9, 154, 169, 220, 229, 244–5
Lewis, Gilbert Newton, 141
light, 54, 138, 169, 177, 187, 195–6
    multiple truths about, 172
    as photons, 148, 195
    ray, 147
lime-water test (for fixed air), 243
logic, 63–7, 169, 244
    bivalent, 192
    many-valued, 171

logical positivism, 6, 58, 205
Lõhkivi, Endla, 207, 217
Longino, Helen, 7
Lowry, T. M., 141, 146
Lynch, Michael P., 165, 174
Lyons, Timothy, 107, 109, 186

Mach, Ernst, 142, 246, 248
Mäki, Uskali, 207
Martí, Genoveva, 97
Marxism, 217
mass, 87, 142
mass–energy conversion, 158
Massimi, Michela, 208, 229–31, 239
Maudlin, Tim, 84, 89, 119, 150, 233
Maxwell, James Clerk, 195
    Maxwell's equations, 183
Maxwell, Nicholas, 20
May, Theresa, 9
McIntyre, Lee, 190
McKenzie, Kerry, 207
McLaughlin, Amy, 214
McLeod, Alexus, 174
meaning, 77, 85–90, 93, 241
meaningfulness, 88–90, 99
measurement, 208, 239
mereology, 91
Merleau-Ponty, Maurice, 61
metaphor, dead, 81
metaphors (various), 43, 80–1, 116, 183, 236, 240
metaphysical realism, 8, 71, 84–5, 91, 150, 205–6, 224, 227, 249
methane, 189
methodology, 63, 244
meter, 100, 177
microreductionism, 157
Miller, Alexander, 73
mind-control, 71–2, 124, 126, 142, 169, 220
mind-framing, 71, 77, 86, 93, 124, 131–41, 220–2, 230
mind-independence, 70–3, 124, 169, 180, 201, 224
    individuative, 72, 132
Misak, Cheryl, 25, 61, 164, 198, 207
Mitchell, Sandra, 239
model-theoretic arguments, 91, 227
    indeterminacy argument, 227
    permutation argument, 227
Mol, Annemarie, 149
monism, 201, 215
    in Kant's philosophy, 232
Moore, G. E., 103, 168
Morgan, Mary, 113
Morrison, Margaret, 113
myth of the given, 74

naming, 95
natural kinds, 244
natural ontological attitude (NOA), 206
naturalistic metaphysics, 130, 149, 162
necessity, practical, 121, 183, 194
Nersessian, Nancy, 113
Neurath, Otto, 6, 25, 70, 205
New York Times, 163
Newtonian mechanics, 87, 142, 147, 170, 172,
    186, 234, 236
  as approximation to relativity, 250
Nguyen, James, 114
Niiniluoto, Ilkka, 6, 69, 71, 82, 87, 100, 132,
    150, 191, 219, 225, 234
Noë, Alva, 49, 61
Norton, John, 134
noumena, 128, 219, 225
noumena vs. phenomena, 74–5, 112
noumenal jam, 82

objectivism, 230
observational reports, acceptance of, 169
ontological principle, 134–6, 139, 161
ontological realism, 73
ontology, 27, 237
  activity-based, 134–6
  change by epistemic iteration, 241
  as an internal matter, 224
  rooted in practice, 221
operational coherence, 4–5, 23–4, 31, 40–8, 126,
    143, 167, 169, 181, 200, 245
  attribution of, 47
  as a matter of degrees, 171
operational conception of reality, 119, 129–31,
    149
operational ideal, 2, 7, 12, 85
operationalism, 3–4, 24, 235, 248
Oppenheim, Paul, 156
orbital, 152, 185
ordinary-language philosophy, 120
Ørsted, Hans Christian, 52
oxygen, 16, 141, 145, 155, 194, 243
  as de-phlogisticated air, 152

Page, Sam, 71, 132
paradigm, 14, 26, 33, 51, 132, 196, 218, 223,
    229, 236–8
  term used by C. I. Lewis, 137
partial realism, 106
part–whole relation, 157
past, argument from the, 73, 123
Pedersen, Nikolaj J. L. L., 165, 175
Peirce, Charles Sanders, 25, 46, 58, 61, 197, 200,
    208, 213–14, 240–1, 250
penguin, 94

perfect model model, 207
perpetual motion, 246
perspectival realism, 222, 229–31
perspectivality, three layers of, 229
perspective, definition of, 229
perspectivism, 81, 208, 237
pessimistic induction, 98, 106, 186–7
Peters, Dean, 106
phenomenology, 61
phlogiston, 151, 153, 155, 194, 233
phlogiston vs. oxygen, 195
photon, 124, 159
Piaget, Jean, 95
Pickering, Andrew, 49, 211, 216
Pihlström, Sami, 129, 133, 220–1, 225
Pirsig, Robert, 23
placebo effect, 121
pluralism, 66, 110, 231–9
  conceptual, 150
  concerning progress, 213–15
  concerning truth, 165, 172, 200
  conservationist, 234–8
  different types of, 237–9
  epistemic, 149, 172, 234
  as equal-opportunity realism, 153
  foliated, 237
  interactive, 238
  and internal realism, 232–4
  ontological, 148–55, 172, 233–4
  and perspectivism, 232
  tolerant, 237
Poincaré, Henri, 107, 177, 246
point, in Euclidean geometry, 88
Polanyi, Michael, 19, 28–30, 50
Popper, Karl, 51, 191
positivism, 69, 204, 248
positron, 121
positron emission tomography (PET), 121
post-truth, 164
practical realism, 217–19, 221
practicality, 5–6, 253
practice, 34, 218, 221, 229, 253
  of perception, 62
  of representing, 120
  semantic, 94
practice-based realism, 217
pragmatic realism, 216, 225
pragmatic scientific realism, 220–1
pragmatic sense-making, 24, 31, 44–8, 56, 134
pragmatism, 4–6, 19, 57–67, 197–203, 252
  as an action-oriented philosophy, 253
  hybrid, 222
  and realism, 216–23
  as relentless empiricism, 60
pragmatist maxim, 58

pragmatist metaphysics, 129–31
pragmatist realism, 206
pre-figuration, fallacy of, 73–80, 90, 95, 104, 112, 116, 217
preservationism, 105–8
preservative realism, 106
Price, Huw, 59, 165, 174, 176, 178
Priestley, Joseph, 152, 194
progress, 8, 117, 125, 247–51
    abundant, 213
    cumulative, 187, 215
    imperative of, 209
    iterative, 208
    pluralist, 236
    pragmatic, 213
    promoted by activist realism, 211–16
    self-improving, 240
    teleological, 213
    various modes of, 213–15
proliferation, 238
promiscuous realism, 126
pronoun, gender-neutral, 21
proper name, 96
property-cluster, 141
    homeostatic, 145
propositional knowledge, 13, 17, 237
    embedded in active knowledge, 20–3
Psillos, Stathis, 98, 102, 106, 108–9, 184
pursuit, 246
Putnam, Hilary, 7, 71, 74, 79, 81, 84, 86, 91, 93, 96, 108, 150, 156, 207, 219–21, 223–9, 232–4

quality, 23
quantum field theory, 97, 99, 104, 148, 154, 159, 195, 234
quantum mechanics, 41, 46, 87, 99, 152, 159, 170, 172, 186–8, 193, 234
    as a closed theory, 238
quantum wavefunction, 128, 147
Quine, W. V. O., 96, 133, 227, 234

Radder, Hans, 61
Ramsey, Frank P., 15, 207
rationalism as anti-realism, 248
rationality, 45, 246
Ray, Greg, 164
real realism, 207, 222
realism. *See also* Scientific realism
    as an activist ideal of inquiry, 209
    basic meaning of, 204
    concerning science, 209
    relation to empiricism, 205
realistic spirit, 1–3, 7, 205–8, 223, 228
reality (real entity), 72, 125, 127, 204

reality (real-ness), 72, 119–24
    achievement of, 142–6
    coherence theory of, 121
    constitutive criterion of, 121
    in degrees, 146
    domain-specificity of, 147
    judgements of, 120
reality, as the object of knowledge, 119
reality, external, 127
reality, formal, 126
reality, in fiction, 126
reclamation of terms, 205
reductionism, 155, 161
reference, 78–100
    causal theories of, 97–8, 233
    as interest-relative, 233
    as internal, 226
reference-borrowing, 93, 97
reference-fixing, 93, 96, 100
reflective equilibrium, 245
Regnault, Victor, 143
Reichenbach, Hans, 76
relativism, 42, 173, 193
relativity
    conceptual, 228, 233
    ontological, 96, 150
relativity, general theory of, 87, 118, 154, 170, 172, 234, 238
relativity, special theory of, 54, 57, 87–8, 99, 137, 177, 187, 234, 246
    pluralist interpretation of, 235
reliance, 121, 183, 194
representation, 111–18
representational activity, 80
Rescher, Nicholas, 128, 232
respect, principle of, 240
rigid body, 99
rigid designator, 98, 100, 233
Rorty, Richard, 74
Ross, Don, 157
Rouse, Joseph, 20, 25, 34, 214, 218
Rowbottom, Darrell, 109
Rumford, Count, 151
Ruphy, Stéphanie, 237
Russell, Bertrand, 17, 201, 203
Ryle, Gilbert, 13, 22

Sagan, Carl, 3
Scharp, Kevin, 165, 175
Schatzki, Theodore, 34
Scheffler, Israel, 232
Schiller, Ferdinand Canning Scott, 62, 198, 202
Schweber, S. S., 171
scientific method, 215

scientific realism, 1, 102, 171, 217, 226, 247–51
  standard, 69, 78, 84–5, 207, 231, 235, 249
  ultimate argument for, 108
Searle, John, 32
second-person interaction, 31, 49
selective realism, 106
self-improvement, 245
Sellars, Wilfrid, 32, 74
semantic move, 121, 170, 180, 203
semantic realism, 219
semantics as a tool for action, 255
sense, 96
Šešelja, Dunja, 246
set theory, 91
simultaneity, 89, 137
single value, principle of, 135, 143
Skulberg, Emilie, 124
Slater, Matthew, 80
Smart, J. J. C., 109
Snowdon, Paul, 23
sociality, 30–3
Soler, Léna, 34–5, 38, 43
Solomon, Miriam, 103
Sosa, Ernest, 19
spandrels, 107
Spurrett, David, 157
squirrel, 58
Staley, Richard, 99
stance, 7
Standard Model, 159
Stanford, P. Kyle, 98, 109
Stang, Nicholas, 220
Steinle, Friedrich, 53
Stoljar, Daniel, 164
Straßer, Christian, 246
structural realism, 107
Stump, David, 133, 244
Suárez, Mauricio, 113
substitution reaction, 189
success, 44, 47, 106, 182, 189, 222, 235
  realist argument from, 108–11, 250
surrealism, 186
Swoyer, Chris, 114
system of practice, 16, 36–9

Tarski, Alfred, 100, 164
Taub, Liba, 43
tautologies (various), 9, 71, 142
tautology, 177
telescope, 64, 222
Teller, Paul, 15, 74, 86–7, 207, 229, 238
temperature, 139, 143–6, 241–2
  negative absolute, 145
  in relation to heat, 144
Thagard, Paul, 24, 42, 45

theory-choice, 14, 196, 236
theory-testing, 181
theory-use, 181, 185
thermometer, 64
Thomson, J. J., 194
tiger vs. shark, 194
Tomczyk, Hannah, 188
Toon, Adam, 112
Torretti, Roberto, 7, 74, 134, 216, 225
transcendental argument, contingent, 134
Trump, Donald, 4, 179
truth, 163–203
  absolutism concerning, 170, 172
  approximate, 171, 191, 250
  constitution of, 166, 170, 180–2
  correspondence theory of, 68, 170, 197–8
  definition vs. criteria of, 203
  in degrees, 87, 171, 185, 190,
    199
  domain-specificity of, 171, 186–7
  empirical adequacy vs., 247
  functioning in practice, 178
  functions of, 165, 168, 170, 174, 199
  in hybrid pragmatism, 222
  as internal, 224
  knowledge vs., 181
  metaphysics vs. epistemology of, 180
  as non-epistemic, 84
  operational, 206
  perspectival, 230
  pluralism concerning, 179–201
  pragmatist theory of, 60, 169, 197
  primary vs. secondary, 165, 178–9, 223
  primary, in empirical domains, 167, 180
  Putnam's notion of, 225
  as a quality, 171, 190–2, 200
  secondary, in empirical domains, 170
  utility vs., 202
  various concepts of, 175
truth functionalism, 174
truth pluralism, 165, 172, 179–201
truth-aptness, 77, 86, 89, 167
truth-by-assertion, 176
truth-by-comparison, 178, 226
truth-by-decree, 177
truth-by-honesty, 177, 179
truth-by-operational-coherence, 167–70,
    175–96, 223, 231, 235, 250
  plurality of, 172, 193–5
Tuomela, Raimo, 32, 74, 82
two tables (Eddington), 154

uncertainty principle, 88, 99
understanding, 44
uniform consequence, principle of, 134

unobservable entities, 1, 69, 146, 153, 205, 246–9
unrestricted inquiry, 236

van Fraassen, Bas, 7, 69, 104, 108–9, 113–14, 125, 247–8
verificationism, 228, 248
verisimilitude, 191
Vickers, Peter, 106–7
Vienna Circle, 6
view-from-nowhere, 117
Vihalemm, Rein, 69, 78, 207, 217–19, 221
vision, 61, 153

water, 93, 97, 116, 155, 195, 238
Waters, C. Kenneth, 7
wave–particle duality, 193, 196
Weber, Max, 36
Weinberg, Steven, 187
Westerblad, Oscar, 45
Wheaton, Bruce R., 194

Whewell, William, 137, 244
whig realism, 103
Williamson, Timothy, 19
Wilson, Mark, 87
Wimsatt, William, xii, 2
winning, 175
Winther, Rasmus, 78
witchcraft objection, 172, 189
Wittgenstein, Ludwig, 16, 18, 26, 30, 70, 86, 111, 166, 168, 207
Woodward, James, 5
world, 70, 73, 75, 79, 84, 92, 104, 116, 127, 219
   external, 80
   multiple meanings of, 128
   ready-made, 74, 91
   versions of, 128, 223
Worrall, John, 107
Wray, K. Brad, 106, 109
Wright, Crispin, 175

Yi, Sang Wook, 241

Printed in the United States
by Baker & Taylor Publisher Services